DATE DUE

OCT 0 7 2009	

ANIMALCULES

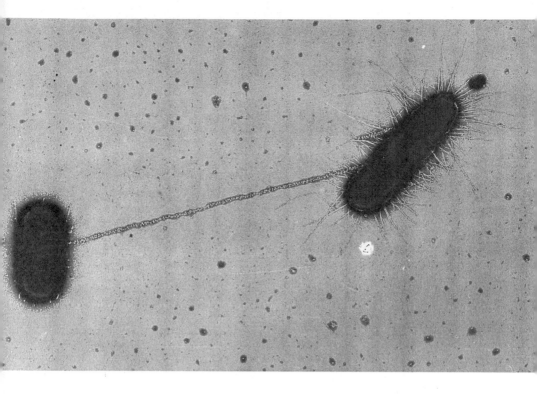

ANIMALCULES

The Activities, Impacts, and
Investigators of Microbes

BERNARD DIXON

**ASM
PRESS**

Washington, DC

Address editorial correspondence to ASM Press, 1752 N St. NW, Washington, DC 20036-2904, USA

Send orders to ASM Press, P.O. Box 605, Herndon, VA 20172, USA
Phone: (800) 546-2416 or (703) 661-1593
Fax: (703) 661-1501
E-mail: books@asmusa.org
Online: estore.asm.org

Library of Congress Cataloging-in-Publication Data

Dixon, Bernard
  Animalcules : the activities, impacts, and investigators of microbes / Bernard Dixon.
    p. ; cm.
  Includes bibliographical references and index.
  ISBN 978-1-55581-500-4 (hardcover)
  1. Microbiology. 2. Microbes. I. American Society for Microbiology. II. Microbe. III. Title. IV. Title: Activities, impacts, and investigators of microbes.
  [DNLM: 1. Bacterial Physiology—Collected Works. 2. Microbiological Techniques—Collected Works. 3. Microbiology—Collected Works. QW 52 D621a 2009]

  QR41.2.D59 2009
  579—dc22

             2008040223

ISBN 978-1-55581-500-4

Cover photo: Evolution under the electron microscope. The electron micrograph shows two *E. coli* cells joined by a sex pilus, through which DNA can travel from one to the other. Gene transfers of this sort have given some strains, such as O157:H7, the capacity to produce toxins causing serious forms of food poisoning (see chapter 11). Reprinted from M. Schaechter, J. L. Ingraham, and F. C. Neidhardt, *Microbe* (ASM Press, Washington, DC, 2006).

Cover and interior design: Naylor Design

Contents

Introduction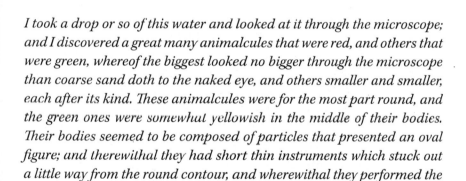

I took a drop or so of this water and looked at it through the microscope; and I discovered a great many animalcules that were red, and others that were green, whereof the biggest looked no bigger through the microscope than coarse sand doth to the naked eye, and others smaller and smaller, each after its kind. These animalcules were for the most part round, and the green ones were somewhat yellowish in the middle of their bodies. Their bodies seemed to be composed of particles that presented an oval figure; and therewithal they had short thin instruments which stuck out a little way from the round contour, and wherewithal they performed the motions of rolling round and going forward; and when they took a rest and fixed themselves to the glass, they looked like a pear with a short stalk.

So wrote the Dutch draper and self-taught microscopist Antony van Leeuwenhoek on 9 February 1702 in a letter to his countryman Hendrik van Bleswyk. Along with an equally remarkable figure, the English polymath Robert Hooke, it was Leeuwenhoek who founded the science of microbiology three centuries ago by constructing and using elementary microscopes to observe otherwise invisible forms of life (see chapter 43). All of today's microbiologists (who, ironically, spend less time observing their subjects directly than studying them in other ways) are indebted to these diligent pioneers. Their work is all the more praiseworthy for being conducted nearly two centuries before Louis Pasteur, Robert Koch, Paul Ehrlich, and their fellow "microbe hunters" began to establish bacteriology as an experimental science closely linked with practical applications for human welfare.

Through the biochemical and serological techniques introduced by these luminaries, recently augmented by those of molecular genetics, our knowledge of what Leeuwenhoek called animalcules has deepened considerably. So too has awareness of our paradoxical relationship with the microbial world. We are conscious that microorganisms continually assail our tissues, those of other animals, and those of plants, and that they exhibit unparalleled versatility and adaptability in doing so. Yet we also recognize that animalcules have fashioned the biosphere and indeed much of the physical world, that they are crucial to the health of our planet, and that they provide many of the means by which we promote human health and environmental well-being.

Our grasp of these twin aspects of microbiology, though formidable, is far from complete—a most enticing factor for young people contemplating a career in science today. On one hand, we confront new, newly recognized, and resurgent pathogens, all posing intellectual and practical challenges. On the other hand, and perhaps even more astonishingly, research in recent years has demonstrated many hitherto unrecognized contributions of animalcules to the ever-renewing, ever-changing texture of nature.

The nitrogen cycle (almost the only aspect of nonmedical microbiology many of us encounter at school) is a prime example. Among its previously unsuspected contributors as nitrogen fixers are *Montastraea cavernosa* in Caribbean coral (7), *Bacillus marisflavi* in plant rhizospheres in the Beijing region of the People's Republic of China (4), and a methanogenic archaeon in deep-sea hydrothermal vents in the Pacific Ocean (8). Even more radically, discoveries such as the anammox reaction (in which bacteria oxidize ammonium anaerobically, using nitrite rather than oxygen as the electron acceptor, producing N_2 gas) have raised profound questions about the validity of the conventional model of the nitrogen cycle.

Meanwhile, advances elsewhere have revealed similar lacunae in our picture of other segments of the cat's cradle of processes that sustain the biosphere. Just two recent examples are the revelation of a fifth pathway of carbon fixation (1) and the realization that picoplankton, tiny unicellular plants, play major roles in mobilizing organic matter deep in the oceans (9).

Sometimes overlooked as an important component in the portfolio of microbiology is the degree to which its insights and practical skills assist other sciences and promote our understanding of the world in often surprising ways. Just over 50 years ago, Jan Kluyver and C. B. van Niel published their splendid book *The Microbe's Contribution to Biology* (6). This reviewed the many spin-offs from studies, largely on bacteria, that were enhancing both knowledge of other aspects of biology and the conceptual foundations of science. Topics ranged from work on phototrophic bacteria, which facilitated understanding of green plant photosynthesis, to emerging ideas about the unity, flexibility, and evolution of terrestrial life.

A 21st-century revision of Kluyver and van Niel would have a good deal more to say on all of these issues in the light of modern molecular genetics. It might, for example, cite the remarkable work of Roman Biek and others in using a fast-evolving organism (feline immunodeficiency virus) to discern the population structure and demographic history of its natural wildlife host, the cougar. As they point out (2), these findings could not have been obtained in any other way and were not apparent from host genetic data.

Another item for inclusion would be the extraordinary paper by Tom Dillehay and colleagues which indicates that when the first Americans came from Asia, they took a coastal route rather than traveling inland (3). The evidence was microbiological: the remains of nine species of marine algae recovered from hearths and other features at an archaeological site in Monte Verde in southern Chile.

Furthermore, there are the ways in which investigations on microbial cilia (shorter cousins of Leeuwenhoek's "short thin instruments" or flagellae) are contributing to our comprehension of another unexpected series of discoveries, in very recent years, on the roles of cilia in human development and disease. For example, patients with polycystic kidney disease, one of the commonest genetic disorders, have cilia whose abnormalities are becoming better understood through research on the flagellae of the alga *Chlamydomonas*.

The present book is based on the "Animalcules" columns which I have been writing regularly since 1996 for *Microbe* (formerly *ASM News*), published by the American Society for Microbiology. I am grateful to Michael Goldberg for asking me to make these contributions and to both him and Patrick Lacey, the production manager of *Microbe*, for their support over the past decade. I also thank the many people who, in recent years, have suggested that the articles could be brought together in book form. Above all, I am grateful to my partner, Kath Adams, for help in manifold ways.

As an edited and orchestrated collection of pieces about animalcules, their activities, and their investigators, this book certainly does not purport to be comprehensive. Moreover, while I have resisted temptations to revise the entire text, or even to modify assertions that seem more questionable with the passage of time, I have added a significant amount of new material here and there in order to accommodate more recent developments. Just two pieces (those on Antony van Leeuwenhoek and Cecil Hoare) did not appear originally as articles in *Microbe*. One piece ("Pioneers of American Microbiology") was published in *ASM News*, though not in the "Animalcules" series.

Contributing to a recent special section of *Science*, James Tiedje and Timothy Donohue (10) pointed out that the incredible diversity of today's microbial world reflects the accumulated evolutionary response to diverse environments over the 3.5 billion years that microorganisms have inhabited Earth. Paul Falkowski and colleagues (5) added that the microbial world drives some of the largest-scale phenomena on the planet, from photosynthesis and nitrogen cycling to pandemics of infectious disease.

Animalcules contains no more than snapshots taken from this vast mosaic of microbial (and macrobial) activity. Experts are unlikely to learn anything on their own subject here, but I hope that they may find items of interest about happenings elsewhere in the jungle.

Bernard Dixon

References

1. **Berg, I. A., D. Kockelkorn, W. Buckel, and G. Fuchs.** 2007. A 3-hydroxypropionate/4-hydroxybutyrate autotrophic carbon dioxide assimilation pathway in archaea. *Science* **318**:1782–1786.

2. **Biek, R., A. J. Drummond, and M. Poss.** 2006. A virus reveals population structure and recent demographic history of its carnivore host. *Science* **311**:538–541.

3. **Dillehay, T. D., C. Ramírez, M. Pino, M. B. Collins, J. Rossen, and J. D. Pino-Navarro.** 2008. Monte Verde: seaweed, food, medicine, and the peopling of South America. *Science* **320**:784–786.

4. **Ding, Y., J. Wang, Y. Liu, and S. Chen.** 2005. Isolation and identification of nitrogen-fixing bacilli from plant rhizospheres in Beijing region. *J. Appl. Microbiol.* **99**:1271–1281.

5. **Falkowski, P. G., T. Fenchel, and E. F. Delong.** 2008. The microbial engines that drive Earth's biogeochemical cycles. *Science* **320**:1034–1039.

6. **Kluyver, J., and C. B. van Niel.** 1956. *The Microbe's Contribution to Biology.* Harvard University Press, Cambridge, MA.

7. **Lesser, M. P., C. H. Mazel, M. Y. Gorbunov, and P. G. Falkowski.** 2004. Discovery of symbiotic nitrogen-fixing cyanobacteria in corals. *Science* **305**:997–1000.

8. **Mehta, M. P., and J. A. Baross.** 2006. Nitrogen fixation at 92°C by a hydrothermal vent archaeon. *Science* **314**:1783–1786.

9. **Richardson, T. L., and G. A. Jackson.** 2007. Small phytoplankton and carbon export from the surface ocean. *Science* **315**:838–840.

10. **Tiedje, J., and T. Donohue.** 2008. Microbes in the energy grid. *Science* **320**:985.

I. Touching Life at Many Points

1 Disseminators Aloft?

"Nine of 75 antibiotic-resistant isolates possessed the int1 gene. Eight of them carried class 1 integrons. One isolate with antibiotic resistance, type ATSSu, carrying an int1 gene, was negative for the presence of integron and gene cassettes." Written in the same city where Gregor Mendel laid the foundations of genetics a century and a half ago, those words would both mystify and delight the man whose patient researches revealed the celebrated 3:1 ratio that first pointed to the existence of particulate inheritance. Mendel began his breeding experiments with the edible pea *(Pisum)* in 1856 at the age of 34. Although no one recognized the significance of this work until the turn of the century, 16 years after he died, Mendel's studies in Brno, Moravia (now in the Czech Republic), were the essential forerunner of today's molecular genetics.

Hence to the research on integrons, gene cassettes, and antibiotic resistance genes being carried out by a group of Mendel's compatriots at the University of Veterinary and Pharmaceutical Sciences in Brno. Their studies would have greatly appealed to Mendel the scientist, but also, for their practical implications, to Mendel the naturalist and gardener.

Monika Dolejska and her colleagues want to know more about the extent to which wild birds act as vectors for *Escherichia coli*, *Campylobacter* spp., salmonellae, and other organisms that can cause human disease. Part of their strategy has focused on the black-headed gull (*Larus ridibundus*), an abundant species in central Europe, where it nests in colonies on water reservoirs. With an estimated 80,000 to 100,000 breeding pairs in the Czech Republic, it seemed at least possible that these gulls could be a significant reservoir for human pathogens, including antibiotic-resistant strains.

As described in the *Journal of Applied Microbiology* (1), the Brno team studied over 250 *E. coli* isolates from cloacal smears of young black-headed gulls from three different nesting sites. Of these, 75 (30%) were insensitive to one or more antimicrobial agents. The major type of resistance was against tetracycline (19% of isolates), followed by ampicillin (12%), streptomycin (9%), and sulfonamides (8%). Nine isolates carried integrons.

Since the nesting colonies examined were all in heavily populated areas with intensive agriculture, these high levels of resistance probably reflect those which occur in both livestock and the human population. The birds visit farmland as they gather food, and they also frequent reservoirs and water sources where wastewater from farms and cities is discharged.

"Our demonstration of the same integrons and inserted gene cassettes in black-headed gull *E. coli* isolates as in human isolates is additional evidence that black-headed gulls were in close contact with human and animal population products, thus coming into contact with bacteria carrying integrons and antibiotic resistance genes," Dolejska and her collaborators wrote. "These are then potentially transmitted to the wild, as well as contaminating drinking water, water for animals and recreational purposes, and feedstuffs for wild animals."

Evidence complementary to that from the Czech Republic was published recently by Nicholas Peak at the University of Kansas, with collaborators there and elsewhere. They set out to quantify the abundance of six tetracycline resistance genes in wastewater lagoons at cattle feedlots, to assess the effects, if any, of different strategies for the use of antibiotics. Their findings, published in *Environmental Microbiology* (3), established a very strong correlation between the level of antibiotic usage and both the abundance and seasonal distribution of resistance genes in the lagoons.

The Brno work is a further component in an emerging picture of the degree, unrecognized in some countries, to which migratory and other wild birds can disseminate both potential pathogens and resistance genes. Newspaper headlines have, of course, highlighted scenarios such as the spread of West Nile virus throughout the United States and the ferrying of influenza viruses between waterfowl, ducks, chickens, pigs, and humans in Southeast Asia. Far from completely understood, however, is the role of free-flying birds as purveyors of enteric bacteria.

Keith Jones and his colleagues at Lancaster University in the United Kingdom have shown that up to 50% of *E. coli* bacteria in coastal seawater can originate from bird droppings. One of their studies revealed that a strain of *E. coli* O157 recovered from gulls and waders in Morecambe Bay on England's northwest coast was identical to that found in cattle on farms in the region.

"The situation with *Campylobacter* is even more dramatic," Jones wrote in *Microbiology Today* (2). "The intestinal environment of birds is favourable for the growth of *Campylobacter* with the result that large numbers are excreted in their droppings."

In Morecambe Bay, the Lancaster group has demonstrated that the seasonal pattern of campylobacters in seawater is similar to that of the main waders, gulls and wildfowl. Jones believes that this finding, together with evidence that the species profile of *Campylobacter* in seawater and mussels is similar to that of birds rather than that of sewage, indicates that birds, not sewage effluent, are responsible for campylobacters in coastal waters.

Mallards appear to be the culprits at another location, a bathing site on the river Lune near Lancaster. They pollute both the freshwater and the surrounding picnic area with the pathogen *Campylobacter jejuni* (in contrast to *Campylobacter lari*, which occurs in the sea but rarely causes disease). "You are more likely to get campylobacteriosis from swimming in freshwater than in seawater," Jones comments.

Another part of Europe showing concern over wild birds as vectors of potentially harmful organisms is Sweden, where researchers at the University of Lund have discerned a considerable range of campylobacters in both geese and shorebirds. Jonas Waldenström and his coworkers took cloacal swabs from lapwings and other species, together with fecal samples from barnacle geese and cattle, in three locations in Gotland during late spring and early summer.

From a total of 247 birds and 71 cattle, they recovered 113 isolates of urease-positive thermophilic campylobacters (UPTCs) and 16 of *C. jejuni.* There were also 18 isolates of *Helicobacter canadensis* and five isolates believed to represent a new genus of microaerophilic, spiral, gram-negative bacteria. The organisms showed an uneven distribution: *H. canadensis* was restricted to geese, while all except one of the UPTC isolates occurred in shorebirds.

Although UPTCs are not considered to be frequent human pathogens, Waldenström and his colleagues said that they might have been underestimated in the past. "The public health implications for the acquisition of campylobacterial disease through direct or indirect contact (e.g., via pollution of open waters) are self-evident," they wrote in the *Journal of Applied Microbiology* (4). "They should be considered in light of the many questions that remain concerning the epidemiology of the most well-studied *Campylobacter* species, as well as species whose prevalence in episodes of undiagnosed human gastroenteritis is presently unknown."

Arguably one of the most vulnerable links in the human food chain for contamination by wild birds is the cultivation of salad and other vegetables. Highlighted at a Society for Applied Microbiology meeting held in Cardiff, Wales, the colonization of fresh produce by *E. coli*, salmonellae, and other potential pathogens is assuming growing importance as more consumers choose "healthier" foods and as sales of ready-to-use vegetables burgeon. The problem is exacerbated by the tenacity with which bacteria (especially salmonellae) adhere to leaf surfaces and the imperfect techniques as yet available to wash them away.

For one reason or another, a picnic by the seaside or river with a lovely salad does not seem quite such a good idea any more.

References

1. **Dolejska, M., A. Cizek, and I. Literak.** 2007. High prevalence of antimicrobial-resistant genes and integrons in *Escherichia coli* isolates from black-headed gulls in the Czech Republic. *J. Appl. Microbiol.* **103:**11–19.

2. **Jones, K.** 2005. Flying hazards: birds and the spread of disease. *Microbiol. Today* **32:**175–178.

3. **Peak, N., C. W. Knapp, R. K. Yang, M. M. Hanfelt, M. S. Smith, D. S. Aga, and D. W. Graham.** 2007. Abundance of six tetracycline resistance genes in wastewater lagoons at cattle feedlots with different antibiotic use strategies. *Environ. Microbiol.* **9:**143–151.

4. **Waldenström, J., S. L. On, R. Ottvall, D. Hasselquist, and B. Olsen.** 2007. Species diversity of campylobacteria in a wild bird community in Sweden. *J. Appl. Microbiol.* **102:**424–432.

2 *Pantoea* and the Locust

"At first we thought that it was smoke from some great fire on the plains, but we soon found that it was a swarm of locusts," wrote Charles Darwin in *The Voyage of the Beagle*. "The main body filled the air from a height of twenty feet to that, as it appeared, of two or three thousand above the ground."

Darwin was particularly impressed by the awesome noise generated by the astronomical numbers of flying insects. Seeking an appropriate simile to describe it, he quoted words from the bible, which portrays locusts as one of the plagues of Egypt: "And the sound of their wings was as the sound of chariots of many horses running to battle."

Could all of this, and the predations caused by rampaging locusts to this day, be attributable to bacteria? That is the implication of some fascinating work being carried out by Rod Dillon and his colleagues at the University of Bath in the United Kingdom. They have been trying to find what precisely drives locusts, which are solitary and sedentary for much of the time, to change so dramatically, forming black clouds to fly huge distances in search of food. Signalling molecules (pheromones crafted from bacterial metabolites in the gut) seem to be an important part of the answer.

The sporadic appearance (and disappearance) of swarms of locusts reflects the fact that the insects have two phases. Different in form, physiology, coloration, and behavior, these are known as solitary and gregarious. When a solitary-phase nymph matures in the presence of many other locusts, it develops into a gregarious adult. If the population density is sufficiently high for sufficiently long, then the locusts form a migratory swarm.

Though smaller and darker than their solitary brethren, swarming locusts (which ride air currents as well as fly) have longer wings. They can travel prodigious distances. Journeys of up to 2,000 miles have been recorded (though not in a single continuous flight), while a black cloud may cover an area of up to 200 square miles at any one time. A swarm of this size contains as many as a million million locusts, with a collective weight of many tons.

In the 1870s, the Rocky Mountain locust and a close relative, the migratory grasshopper, destroyed many prairie farms in the United States and Canada. At around the same time, swarms of desert locusts reached England, probably from West Africa. The last great pestilence caused by the African locust was in Ethiopia in 1958, when the rampaging insects destroyed 167,000 tons of cereals, enough to feed a million people for a year. Even today, the damage caused by swarms can be immense, despite the use of chemical sprays and the very recent introduction of biological control with the fungus *Metarhizium flavoride*.

Clearly, greater understanding of what triggers the shift from the solitary mode to the gregarious, aggressive mode could be both instructive and useful. The last 20 years have seen an emerging consensus that several types of stimuli, including tactile and visual clues, contribute to the process of "gregarization." Particularly significant is chemical communication in the form of pheromones: volatile substances, effective in vanishingly tiny concentrations, which in other insects perform functions such as attracting mates and promoting social cohesion.

During the 1990s, several research groups detected aromatic compounds in the atmosphere surrounding gregarious locusts and their fecal pellets and found that they played an essential role in gregarization.

The predominant pheromones released by feces from both adults and juveniles were guaiacol and phenol.

The source of these compounds remained a mystery, however, until the investigation by Rod Dillon and his coworkers. They first reared locusts in the conventional manner and confirmed that volatiles from adult and juvenile fecal pellets had profiles, including guaiacol and phenol, similar to those reported by previous workers. Next, in order to explore the possible involvement of the gut flora in the production of the pheromones, they sterilized eggs (on the surface) and reared the locusts in a sterile isolator system. Using as food grassland bran that they had gamma irradiated and freeze-dried, they were able to establish a breeding colony of axenic locusts.

Compared with fecal pellets from conventional adult and juvenile locusts, those from the germ-free insects turned out to contain a considerably lower level of phenol and no guaiacol at all. Dillon's next step was to see whether the introduction of *Pantoea agglomerans*, a bacterium commonly isolated from locusts, restored synthesis of the two phenolics. So it proved. When the organism became established in the gut, guaiacol and phenol reappeared in the feces.

The University of Bath researchers also used fecal pellets from the axenic insects as a culture medium to grow this and other bacteria associated with locusts. Their aim was to determine their efficacy in producing the phenolics in vitro. As reported in the *Journal of Applied Microbiology* (1), the inoculation of germ-free pellets with *P. agglomerans* led to the synthesis of large quantities of guaiacol and smaller amounts of phenol.

Two other members of the *Enterobacteriaceae* obtained from the locust gut, *Klebsiella pneumoniae* subsp. *pneumoniae* and *Enterobacter cloacae*, also synthesized guaiacol from axenic pellets. *Enterococcus casseliflavus*, an enterococcus representative of the many different species found in the locust gut, yielded small quantities of guaiacol but no phenol. A strain of *Serratia marcescens* that is pathogenic in locusts produced neither of the two phenolics.

Another set of experiments, with conventional (nonaxenic) locusts, established that diet strongly influenced guaiacol synthesis, which was considerably higher when the insects were given fresh wheat seedlings than when they were given freeze-dried gamma-irradiated grass (the same as that fed to the germ-free locusts). Guaiacol production was very low when either diet was incubated with *K. pneumoniae* subsp. *pneumoniae* or *E. casseliflavus*. Clearly, digestion of the plant material in the gut was a prerequisite for bacterial formation of guaiacol.

Indeed, the most likely precursor of the phenolics is vanillic acid, derived from the lignin of plants eaten by the locusts. Fecal pellets from both conventional and axenic locusts contain large quantities of vanillic acid, which can be converted to guaiacol by simple decarboxylation.

It seems that the enzymes responsible for nonoxidative decarboxylation of aromatic acids, as reported in other organisms, may have an adaptive role in the life of the locust. This would not be the first example of such an activity by indigenous gut bacteria, which are presumably acquired serendipitously in food. Previous papers from Dillon's group showed that the production of antimicrobial phenolics is important in defense against infection.

"It is suggested that the insect hindgut, where the fecal pellets are produced, is an environment which is selecting for microorganisms capable of firstly withstanding, and secondly catabolising, aromatic compounds," the authors concluded. In a moderately mutualistic association, the bacteria adapt to metabolize the remnants of plant digestion into both pheromonal and antimicrobial compounds. "This dual benefit for the insect suggests a closer degree of integration between the locust and its microbial community than was previously suspected."

In her book *Animals Can Do Anything*, Jean George described the transformation of sluggish, solitary locusts into rampant swarms as one of the most remarkable phenomena in the whole of the natural world (2). "The transformation can be induced by entomologists simply by putting a locust in its solitary phase into a cage with others of its kind," she wrote. "Crowded together, the sight of the others stimulates a hormone that shrivels them all into tiny monsters."

Over 3 decades later, we recognize that things are rather more complex, involving not only visual, but also tactile and pheromonal signals. Jean George would have been astonished to learn that a crucial role in this extraordinary process is played by bacteria in the locust gut.

References

1. **Dillon, R. J., C. T. Vennard, and A. K. Charnley.** 2002. A note: gut bacteria produce components of a locust cohesion pheromone. *J. Appl. Microbiol.* **92:**759–763.

2. **George, J.** 1973. *Animals Can Do Anything.* Souvenir Press, London, United Kingdom.

3 The Microbiology of Art

In thinking about microbiology and cultural heritage, the first, second, third, and possibly only things that come to mind are deterioration, destruction, and devaluation. Microorganisms are so versatile in their enzymatic repertoires that few, if any, of the materials and pigments used by sculptors and painters are invulnerable to attack. It was a microbiologist, too, who first discovered that the unsightly brown marks found on old prints and books were a consequence of fungal infection. Working at the University of Kent in Canterbury, United Kingdom, Guy Meynell popped some fragments of affected paper under a microscope and found that a phenomenon long known in the antiquarian book trade as "foxing," and misattributed to chemical impurities, was really a manifestation of fungal mycelium (see chapter 10).

Alongside such evidence of microbial attack, recent years have seen the emergence of a new, artistic branch of applied microbiology. Animalcules are now being harnessed to rescue and restore our cultural heritage. One of the most remarkable examples, developed by a leading pioneer of the new approach, Giancarlo Ranalli, is the restoration of ancient frescoes. These are paintings done with watercolors on freshly spread moist lime plaster, often in churches. Many have suffered woefully over the decades and centuries from various forms of biodegradation.

Based at the University of Molise in Campobasso, Italy, Ranalli has combined whole bacterial cells with a purified bacterial enzyme to restore a fresco by Spinello Artino, a 14th-century Italian painter of the Sienese school. Ironically, this was an instance in which much of the damage had been caused by earlier, unsuccessful attempts at restoration. The fresco is the *Conversione di S. Efisio e battaglia (The Conversion of Saint Efisio and Battle),* which measures 3.5 by 7.8 m and is located in the Monumental Cemetery in Pisa.

Along with several other frescoes by painters such as Antonio Bonaiuti and Benozzo Gozzoli, it was badly damaged by a bomb that fell in 1944. During the 1980s, technicians attempted to preserve the precious artifact by sticking gauze to the intelaggio (front surface) with animal glue. As reinforcement for the fragile structure, they then used casein to apply canvas cloth to the reverse side and to fix the fresco to a supportive sheet of asbestos.

One of the greatest challenges faced by Ranalli and his coworkers was how to detach the gauze from the previous restoration efforts that was still adhering to the layer of paint on the intelaggio. Soaking it with solvents did not solve the problem. "Over the years the glue had altered greatly, and had become very hard and resistant to commonly used solvents," Ranalli explained. "Our task was to dissolve the animal glue on the paintwork itself to release the gauze residues without removing the casein supporting the back of the fresco."

As reported in the *Journal of Applied Microbiology* (1), the answer proved to be microbiological. First, however, the University of Molise team had to determine the precise nature of the unwanted organic matter on the fresco, some of which had been made more resistant by being polymerized by substances used in earlier restoration attempts. It was also important to ensure that any biorestoration work would not harm inorganic pigments, which would be unaffected by biorestoration.

A variety of preliminary experiments demonstrated that the optimal procedure to solve the problem would be to apply appropriate whole bacterial cells to degrade the tenacious organic matter, freeing the gauze from the fresco, and then to add purified enzyme to attack the residues released. The specific agent selected was *Pseudomonas stutzeri* (strain

A29), applied as a layer of cotton wool soaked in bacterial culture, and a powerful protease. Temperature had a marked effect on the process. At 28°C, the cellular activity was so intense that treated areas were clean and free of the animal glue, with no structural damage. Organic residues were then quickly and efficiently removed by the protease.

Ranalli attributed the success of his work to the combination of constitutive and inducible enzymes in the *P. stutzeri* repertoire. "The use of microorganisms is more effective than just a single enzyme that attacks only specific linkages. The greatest advantage of bacterial/enzymatic biorestoration, compared with traditional methods (chemical, physical and mechanical), is that this new method is not destructive and removes only extraneous substances or altered compounds in the fresco."

Giancarlo Ranalli began to work on frescoes after tackling another aspect of the deterioration of cultural heritage. Throughout the world, atmospheric pollution is threatening, and in some cases destroying, historic monuments and buildings. Although natural processes are partly responsible, much of the blame attaches to acids released into the air as a result of industrial activities. Sulfur dioxide is a prime culprit, dissolving in water to produce sulfurous acid that dissolves limestone. Further corrosion can be traced to nitrous oxide from automobile emissions and other sources, which reacts with water to form nitric and nitrous acids.

By using appropriate microorganisms, Ranalli and his colleagues achieved several successes in removing sulfates, nitrates, and organic substances from artistic stonework. They believe that bioremediation has a major role to play as a "soft" technology in restoring monuments and other artifacts degraded by atmospheric pollutants.

However, microorganisms play a negative role in this context, too. Eberhard Bock and colleagues at the Institute for General Botany and Microbiology in Hamburg, Germany, have shown that air pollution is not the whole story. Bacteria are also heavily implicated in the corrosive destruction of monuments and buildings. Even more unexpected is the identity of the organisms responsible for the damage.

The first clues emerged when Bock used electron microscopy to study what was happening on and inside exposed concrete and sandstone. Bacteria, subsequently identified as nitrifiers, were present in considerable numbers. Scrutiny of the stonework of Cologne and Regensburg cathedrals revealed *Nitrosomonas, Nitrobacter*, and several other types of bacteria that hasten the destructive process.

One reason why this type of corrosion was not recognized much earlier is that the organisms work largely inside the stone. They are probably deposited on the surface in dust particles before rain or a thunderstorm. Being very sensitive to light, they begin to thrive only after being carried beneath the surface by moisture. The sandstone of the cathedrals was heavily contaminated with nitrifiers to a depth of 5 mm.

Their astonishing power is illustrated by experiments in which the Hamburg researchers inoculated concrete blocks with *Nitrobacter* and *Nitrosomonas* isolated from the cooling tower of a power station near Cologne. The bacteria generated about 14 ml of concentrated nitric acid per concrete block per year, leaving the original surface totally corroded. The acid had converted the binding material of the concrete into calcium nitrate, yet the organisms themselves were able to tolerate the acid.

In 1999, the world's first international meeting on microbes and art took place, appropriately enough in the beautiful Italian city of Florence. Although the conference theme was microorganisms in the degradation and protection of cultural heritage, only 5 out of the 27 papers were devoted to biorestoration. Most of the remainder catalogued the grim details of damage being caused day by day, year by year, to monuments, statues, frescoes, paintings, and other artifacts of our cultural heritage. Thanks to Giancarlo Ranalli's work in the laboratory, in the Pisa Monumental Cemetery, and elsewhere, the next event of this sort may be able to focus greater attention on amelioration than on deterioration.

Reference

1. **Ranalli, G., G. Alfano, C. Belli, G. Lustrato, M. P. Colombini, I. Bonaduce, E. Zanardini, P. Abbruscato, F. Cappitelli, and C. Sorlini.** 2004. Biotechnology applied to cultural heritage: biorestoration of frescoes using viable bacterial cells and enzymes. *J. Appl. Microbiol.* **98:**73–83.

4 Why Do They Do It?

Your name is *Buellia frigida.* You are a lichen, living in and on a rocky outcrop, part of a dry valley in the Antarctic. With little water, sparse nutrients, and temperatures swinging by 40°C during the course of the day, your home is one of the most inhospitable places on the entire planet. Over a whole year, the temperature ranges as much as 60 to 70°C. You adhere tenaciously to the rock, suffering frequent desiccation, since your only water comes in the form of snow, most of which rapidly sublimes back into the atmosphere. Desperately short of nourishment, you cling to life itself.

So why bother about sex? Many other inhabitants of the earth, based in biologically much richer regions, reproduce asexually or vegetatively. Why then should a seemingly primitive organism, existing in extremis, waste precious resources of materials and energy on the paraphernalia of sex?

The question has attracted the attention of lichenologists at the University of Nottingham in Nottingham, United Kingdom, and they are beginning to get some answers. Not only that; what they have found leads them to suggest that some aspects of lichen behavior might be considered "quite scandalous."

These allegations emerge directly from the nature of lichens as symbiotic associations between a "mycobiont" (a fungus) and a "photobiont" (a green alga and/or a cyanobacterium). Photobionts can live independently, but partnership is mandatory for a mycobiont to complete its life cycle. Although lichens proliferate exceedingly slowly, their symbiosis is an astonishingly successful arrangement. Some thalli endure for hundreds and possibly thousands of years, despite, and possibly because of, their prodigious indulgence in sex.

Fabian Seymour, Peter Crittenden, and Paul Dyer were struck by the apparent incongruity of lichens surviving in a variety of hostile environments yet producing sexual structures, often in great profusion. They do so not only in the Antarctic and Arctic, but also in deserts where temperatures reach 55°C during the daytime but plunge to below freezing at night. There are, for example, spectacular yellow displays of *Teloschistes capensis* in the Namib Desert in Namibia, southwest Africa.

Another reason why the occurrence of sexuality in lichens is puzzling is that they can spawn offspring quite satisfactorily by nonsexual means. Lichens make vegetative "isidia" and "soredia," containing both fungal and photosynthetic components, which are dispersed by wind, rain, and small animals. In contrast to the sexual process, this means that the two partners are present to establish a new thallus when the vegetative propagule arrives in a new location. The sexual counterparts, ascospores, carry only the fungus (usually a member of the *Ascomycotina*). A chance encounter with a compatible photobiont is thus necessary to initiate a further symbiotic partnership.

Lichen-forming ascomycetes breed in one of two ways. In heterothallism, the fungus is self-sterile and therefore requires a compatible partner of a different mating type. In homothallism, the fungus is self-fertile but may retain the capacity to consort with similar genetic strains. While this much has been known for many years, technical difficulties have held back deeper understanding. The main problems have been those of studying organisms that reproduce at an abysmally low rate and of encouraging them to adopt in the laboratory the same sexual habits which they have brought to a fine art in hot deserts and frozen wastes.

The Nottingham researchers used molecular markers to establish clearly the mode of reproduction in two lichen-forming fungi, *Graphis scripta* and *Ochrolechia parella*. They generated random amplified polymorphic DNA fingerprint markers for sets of single-spore progeny isolated from one apothecium. The resulting fingerprint patterns within sets of progeny proved to be uniform, demonstrating homothallic reproduction in both fungi. As reported in *Mycologist* (1), these researchers also investigated three species of *Cladonia* found in contrasting environments. This time, spores from the same apothecium were not genetically uniform, indicating heterothallic breeding in all three species.

Why, then, does sexuality thrive in microbial consortia inhabiting some of the toughest environmental niches on the planet? The Nottingham team suggested three possible explanations. First, dissemination. "Sexual spores are generally smaller than vegetative structures such as isidia and soredia, so are likely to be dispersed over longer distances," they wrote. "They can also be produced in larger numbers with less metabolic investment per propagule. . . . Sexual spores may be preferred because they are more resistant to adverse environmental conditions, and can be stored safely within ascomata whilst awaiting conditions favourable for dispersal and germination."

Another reason for sexuality, certainly the type involving heterothallism and, to a lesser degree, homothallism, may be to create genetic novelty. "Little variation, associated with extensive clonal development and/or inbreeding, might place lichen populations at risk if a particularly widespread genotype was unable to withstand altered climatic conditions or compete with new strains of co-occurring organisms." Finally, lichens are vulnerable to attack by parasitic fungi. They may therefore have evolved sexuality and spore dispersal (involving only the mycobiont) as a means of escape.

The need of mycobionts to find new photosynthetic partners was what led the Nottingham group to describe the lichen world in terms which, within the scholarly literature, smack of the lurid. It is this requirement that explains why these microscopic symbionts are so sexually prolific. They have to fashion and release large numbers of ascospores so that, when just a few of them land in a suitable place, they have a good chance of finding appropriate partners to consummate their relationships.

But there is more, in the form of "scandalous interactions." Observations of germinating spores of *Xanthoria parietina* revealed that it can form a casual liaison with an alga not specific to that fungal species. "This photobiont could then be replaced by a species-specific partner at a later stage by 'theft' from nearby sorelia or even other lichen thalli," Seymour and his colleagues wrote. Although some lichenologists are skeptical about the idea, it is possible that fungal spores often descend upon existing lichens and usurp the function of the resident mycobionts.

Unlike other ascomycetes, those involved in lichen partnerships produce ascomata that persist for several years, generating countless ascospores and indulging in "a positive orgy of sex." Meanwhile, because the fruiting bodies are vulnerable to parasitic invaders, they synthesize chemical weapons to protect the thallus. Work at the University of Nottingham has indicated that some of these bioactive substances have pharmacological activities with potential value in human medicine.

In the past, one lichen product, usnic acid, was marketed in Germany, Finland, and Russia for the treatment of athlete's foot, ringworm, lupus, and other skin conditions. Another, obtained from *Roccella*, has long been familiar to scientists as the litmus of indicator paper. The research by the group at the Univeristy of Nottingham points to a new and similarly diverse range of applications. The relevant genes will, of course, need to be cloned and expressed in a suitable vector. Biotechnologists are unlikely to be thrilled by the prospect of cultivating organisms whose working lives are measured not in hours or days, but in years or longer periods.

Xunlai Yuan at the Nanjing Institute of Geology and Paleontology in Nanjing, People's Republic of China, together with colleagues at the University of Kansas, Lawrence, and Virginia Polytechnic Institute and State University, Blacksburg, has presented evidence (2) that the lichen style of symbiosis is some 600 million years old. They found lichen-like fossils in marine phosphorite of the Doushantuo Formation at Weng'an in the southern People's Republic of China predating even the emergence of vascular plants on Earth. Sexuality in extreme environments has apparently been extraordinarily persistent. Microbiologists may be puzzled, but it clearly has formidable merits.

References

1. **Seymour, F. A., P. D. Crittenden, and P. S. Dyer.** 2005. Sex in the extremes: lichen-forming fungi. *Mycologist* **19:**51–58.

2. **Yuan, X., S. Xiao, and T. N. Taylor.** 2005. Lichen-like symbiosis 600 million years ago. *Science* **308:**1017–1020.

5 Out of the Blue

I was talking recently to a medical microbiologist who, though well aware that there are beneficial as well as harmful bacteria, categorized viruses solely as human, animal, and plant pathogens. His view is hardly surprising, since many books still portray all viruses as causes of death and disease, whether they are historic enemies, such as the agents of smallpox and poliomyelitis, or the "hot" viruses responsible for Lassa fever and similar infections today. Just occasionally, accounts of their predations are offset by a brief expression of appreciation for the viruses responsible for the pleasant, variegated colors of certain varieties of tulip.

From a biological standpoint, it has always seemed strange that by far the most numerous biological entities on Earth should be totally inimical to human and other forms of life. Modern molecular studies have underlined that implausibility. And now, a century after researchers began to wrestle with the nature and identities of "filterable viruses," we are beginning to discern the truth. In reality, viruses appear to play extensive, crucial roles in the biosphere, to our benefit and that of other inhabitants of the planet (see chapter 17).

One package of research that has deepened our comprehension of terrestrial viral life is the United Kingdom's Marine and Freshwater Microbial Biodiversity (MFMB) program. With £7 million in funding from the Natural Environment Research Council, it has supported more than 50 projects over the past 5 years in laboratories throughout the country and at sites at home and overseas. In June 2005, participants met in London to review the outcomes, drawn from aquatic sites as diverse as the Arabian Sea, the deep Pacific, a beach in the Welsh town of Aberystwyth, and a small pond in northern England's Lake District.

One of the MFMB discoveries concerned the microalga *Emiliana huxleyi*, studied by researchers at Plymouth Marine Laboratory. They reported strong evidence that a previously unknown virus is responsible for terminating blooms of the alga. It was present in a dying bloom in the English Channel but turned out to be comparatively common. Together with a group at the University of East Anglia, Norwich, United Kingdom, the Plymouth team also found that virus-induced bloom crashes release dimethyl sulfide (DMS). As well as being important in the global sulfur cycle, DMS can form a sulfate aerosol that acts as a seed for cloud droplets. In other words, a marine virus seems to exert a cooling influence on the climate.

Meanwhile, work at the University of Warwick has revealed an intriguing instance of symbiosis between viruses and cyanobacteria in the oceans. Phages infecting cyanobacteria clearly perpetuate themselves by harnessing energy from photosynthesis by the blue-green algae they infect. What has now emerged is that the viral genome contains genes coding for key elements of this photosynthetic machinery. The virus actually helps the alga to keep photosynthesis going longer than would be possible in noninfected cells. Since there can be up to 10 million of these phages in a milliliter of seawater, a significant proportion of the oxygen we and other animals breathe may be a product of virus infection.

Other hitherto-unknown microorganisms recovered through the MFMB program are bacteria that break down methyl bromide. Formed by burning biomass and by cars using gasoline with lead additives (as well as being used as a soil fumigant in some countries), methyl bromide destroys about a quarter of all of the ozone lost from the stratosphere. Plymouth and Warwick investigators have now cultivated three novel types of bacteria that degrade the gas and have used molecular probes to show that the oceans contain many others with this capability.

A more synoptic focus of the Natural Environment Research Council-sponsored program was biodiversity. Have we underestimated the richness of terrestrial microbial life? Or have we been guilty of the reverse error, erroneously extrapolating and multiplying the results of local surveys to a global level?

Studies at a site in England and one in Denmark, together with an analysis of worldwide data, indicate that the former conclusion is correct. The work was carried out by researchers at the University of Oxford and the Centre for Ecology and Hydrology. The first location was a 1-hectare freshwater pond in the Lake District called Priest Pot (which, as the subject of over 130 scientific papers, is probably the most intensively studied pond in the world). The second site, surveyed in association with workers from the University of Copenhagen, was Niva Bay in Denmark.

A full taxonomic inventory for all eukaryotic organisms revealed 1,278 different groups at Priest Pot, including new species of protozoa, and 785 at Niva Bay. However, when the researchers reviewed these findings alongside samples and data from elsewhere around the world, they reached two superficially contradictory conclusions. First, several groups of nonmarine protozoa are indeed cosmopolitan. Second, many organisms whose morphology and behavior have led to their being categorized as the same species are in fact genetically dissimilar.

One example is the ciliate *Cyclidium glaucoma*, which occurs on every continent, looking the same and tolerating a wide range of salinities. Populations recovered from similar habitats were genetically similar, even if widely separated. Those in different habitats were genetically dissimilar, even if geographically close. Thus, isolates from highly salty ponds in Spain and the Great Salt Lake in the United States have identical rRNA gene sequences. There are other genetic matches between populations from brackish waters in Spain and Antarctica and between those from freshwater sites in the United Kingdom and Thailand.

However, genetic analysis of zooflagellates showed that conventional taxonomy grossly underestimates the extent of biodiversity. Not only does the phylum Cercozoa probably contain several thousands of species, but relatively well-studied "species" may include as much diversity as entire orders of higher organisms. One supposed species could contain lineages that diverged hundreds of millions of years ago.

The most striking of a series of discoveries during the 5 years of MFMB was the recognition of an entire new division of bacteria. Designated

JS1 because investigators at Bristol and Cardiff Universities recovered the first organisms from Japan Sea sediments, members of the division were later found to be widespread in marine sediments. Though none have yet been cultured, sequence analysis suggests that they metabolize carbon substrates by using energy obtained from the reduction of sulfates in anaerobic sediments.

Similar work by teams from the Universities of Newcastle upon Tyne and Kent identified new *Actinobacteria* and indicated that over 1,000 other species remain to be discovered. Moreover, the use of dispersion differential centrifugation and selective culture techniques can aid the recovery of a wide range of actinomycetes, even from the extreme depths of the Marianas Trench. One potentially momentous finding was the identification of a new marine species of *Verrucosispora*. It produces a chemically unique antibiotic, abyssomicin C, that inhibits the growth of methicillin- and vancomycin-resistant strains of *Staphylococcus aureus*. The same teams also located organisms with novel nitrile hydratases, which may have applications in biotechnology.

Looking to the future, the MFMB program generated both novel theoretical approaches and new practical tools to facilitate bioprospecting for as-yet-unknown microorganisms in the sea. The Newcastle group evolved innovative computer software that visualizes bacterial taxonomy in three-dimensional space, revealing gaps where unrecognized organisms may exist, and the team in Cardiff developed an isosampler capable of collecting living material at pressures up to 670 atm and temperatures over 100°C.

6 Reflections on Cellulolysis

Henry Tribe of Wolfson College, Cambridge, United Kingdom, was lifting a cardboard box containing a bottle of Turkish liqueur called Yeni Rakı from his minicellar when the bottle fell through the bottom of the box. He was relieved not to lose any of the contents. Nevertheless, Tribe was surprised when he discovered that the lower side of the box had been colonized and degraded by a fungus, although the cardboard was very strong and had a glossy, water-repellent surface. This was obviously an aggressive organism.

Being a mycologist, Tribe decided to investigate and quickly found that a low-temperature fungus was to blame. As he and Roland Weber reported in *Mycologist* (4), "the interior of the bottom part of the box was softened and covered with the ascomata of *Myxotrichum chartarum*. These were massed together in an essentially pure stand which was so dense that it formed a continuous chocolate-brown crust accompanied by very little aerial mycelium."

While this was the first report of *M. chartarum* attacking such a tough target, the organism was discovered and named as long ago as 1832. Christian Nees von Esenbeck, a German botanist who devoted most of his life to the study of fungi and algae, found it on writing paper and described its capacity to degrade the paper as a substrate. Its appearance in Henry Tribe's wine cellar was no doubt associated with the fact that his so-called minicellar was in fact simply the space under the floorboards. Other authors have reported finding the organism in rich organic soil, plant debris, and dung from herbivores.

Although intolerant of high temperatures, *M. chartarum* can proliferate slowly at 5 to 7°C, probably close to the temperature in the aerated underfloor space where it was well established. The fungus was subsequently recovered from another of Tribe's cardboard boxes, this one containing a bottle of Israeli Sabra.

Clearly, paper/cardboard is a prime substrate for *M. chartarum*. One survey of the fungal order to which it belongs, the Onygenales, showed that paper was by far its most favored source of carbon (3). Further work on Tribe's isolate, cultured on various simple media, has shown that the best of these is mineral salts agar with a sheet of cellophane as the only available carbon.

Because cellophane is an ideal carbon source for cellulolytic fungi, Tribe and Weber included in their report a reminder of the need for appropriate precautions when it is used as a drying film for electrophoresis gels and as a seal for glass jars containing jams and other preserved foods. The material should first be immersed in a jar of distilled water and autoclaved thoroughly to remove any plasticizers.

I did not have a wine cellar, and I knew nothing whatever about Onygenales or *M. chartarum* when I did my own first, frustrating experiment on soil organisms and paper over 50 years ago. However, I was keenly interested in microbiology, and one day I learned from a book addressed to "boy scientists" about what seemed to be an enticing experiment.

The author advised readers to tear a small piece out of a newspaper, lay it on some damp soil in a plant pot, and keep it moist and warm for several weeks. The paper would gradually disintegrate, he said, and eventually disappear completely as it was attacked by animalcules from the soil. These would probably be the myxobacterium *Cytophaga*, together with *Cellvibrio* and other cellulolytic microorganisms. We were not to worry our young heads with complexities of that sort, however, but simply watch what happened.

Following the instructions diligently, I observed the experiment every day, but nothing happened. Even with different newspapers and different soils, the paper remained disappointingly resistant to biodegradation. So I asked my science teacher. "You've made the mistake of using glossy paper," he said. No, I hadn't. "Then you have let the soil dry out." No, I hadn't. "You must be doing something wrong," the teacher concluded angrily. "Think for a moment. If microorganisms in the soil did not break down paper, we would be ankle deep in the stuff by now. Anyway, I'm busy. Go away."

Not the most helpful encounter, you may feel, with a young lad keen to learn about the natural world. I was certainly dismayed at the time. The situation was retrieved, however, several months later when I came across virtually the same experiment in another book, but this time with scraps of filter paper instead of newsprint.

This time it worked a treat. In less than 2 weeks, the piece of Whatman no. 1 was virtually invisible. Why the difference? The answer came from Stephen Cummings and Colin Stewart at the Rowett Research Institute in Aberdeen, Scotland (2). It emerged from their study of the degradation of municipal solid waste, 28 million tonnes of which are generated in the United Kingdom every year. Much of this finds its way into landfill sites in remote parts of the country.

Just under a third of the material consists of paper and cardboard, and the Rowett researchers were trying to determine the effectiveness of various landfill microorganisms in breaking down newspaper, in particular. They worked with four cellulolytic bacteria (two species of *Eubacterium* and two of *Clostridium*) isolated from a landfill site. The organisms attacked cellulose robustly when inoculated and grown anaerobically on Whatman no. 1 filter paper, which consists of comparatively refined cellulose, but the bacteria performed far less spectacularly when they were presented with newsprint, generously provided by Aberdeen Journals Ltd., publishers of the *Aberdeen Evening Express*.

Even the most powerful strains, which were extremely active when attacking a diet of filter paper, failed to rise beyond the mediocre when offered Scottish newsprint instead. Part of the explanation for the disparity was the presence of printing ink in the paper. Although not directly toxic to the bacteria, the ink masked the surface of the paper, stopping the organisms from adhering to the cellulose fibers.

The main problem, however, was that as much as 24% of the newspaper consisted of lignin, a polymer of high molecular weight. This prevented the microorganisms from using their cellulases to tear apart cellulose and other inherently susceptible polymers.

Using gas chromatography, electron microscopy, and spectrophotometry, Cummings and Stewart had demonstrated, under anaerobic conditions, something which I had discovered by a considerably more simplistic aerobic approach half a century previously. Both experiments are in accord with a third report (1) of newspapers deeply buried in a landfill site that were still legible when they were recovered 25 years later.

Sadly, my unhelpful schoolteacher is long deceased, so I have not been able to inform him of this work or of Henry Tribe's wine cellar and *M. chartarum*. Nor, come to think of it, can I bring the same man up to date with a whole series of topics upon which he regaled science classes with what even then appeared to be imprudent certitude. He insisted, for example, that gaps would always be essential in railroad tracks, allowing them to expand on warm days, yet all over the world today trains run quite safely on continuous welded rails.

Likewise, we were assured that it would never be practicable to build skyscrapers in London like those of New York and other American cities because of the soft underlying London clay. Not so, as we now know. So, too, with my science teacher's calculations that established the sheer impossibility of a spacecraft attaining the escape velocity required to leave the Earth.

Given this trio of erroneous assurances about big technology, it is hardly surprising that he was wrong about animalcules.

References

1. **Booth, E. J.** 1965. Buried 25 years and still legible. *Am. City* **80:**26–34.

2. **Cummings, S. P., and C. S. Stewart.** 1994. Newspaper as a substrate for cellulolytic landfill bacteria. *J. Appl. Microbiol.* **76:**196–202.

3. **Currah, R. S.** 1985. Taxonomy of the Onygenales: Arthrodermataceae, Gymnoascaceae, Myxotrichaceae and Onygenaceae. *Mycotaxon* **24:**1–216.

4. **Tribe, H. T., and R. W. S. Weber.** 2002. A low-temperature fungus from cardboard, *Myxotrichum chartarum. Mycologist* **16:**3–5.

7 Jelly from Space?

Many microbiologists are vaguely aware of occasional media reports about the gelatinous material, sometimes called *pwdre ser,* which is said to arrive on Earth during meteor showers. Few will know the reality behind these stories. Some may have been berated, as I have, by the taunts of true believers, who insist that orthodox science is fecklessly ignoring evidence of fungal arrivals from space.

Most will be surprised to find that the folklore surrounding phenomena such as star jelly now forms part of a formal discipline known as ethnomycology. Indeed, Ángel Nieves-Rivera and Andrew White, writing in *Mycologist* (2), asserted that the beliefs found in ancestral cultures concerning links between space phenomena and terrestrial blobs of jelly are "useful." In their view, such ideas enhance ethnomycology because they illustrate the "range of complexity and conditions in which a fungus myth was developed."

Over the centuries, there have certainly been lots of descriptions of weird gelatinous goo observed shortly after the appearance of shooting stars (meteors that burn up when they enter the atmosphere). The goo is often brightly colored, from yellow to red, and is sometimes said to stink unpleasantly.

The British Isles, unsurprisingly perhaps, in light of their often damp weather, have been the source of many of these tales. Probably the earliest came from the pen of the English poet Sir John Suckling in 1641:

As he whose quicker eye doth trace
A false star shot to a mark'd place
Do's run apace
And, thinking it to catch,
A jelly up do snatch.

Just over a decade later, the English philosopher Henry More wrote "that the Starres eat . . . that those falling Starres, as some call them, which are found on the earth in the form of a trembling jelly, are their excrement." A few years later another English poet, John Dryden, described how "when I had taken up what I supposed a fallen star I found I had been cozened with a jelly." The Scottish novelist Sir Walter Scott, in *The Talisman* (1825), wrote, "Seek a fallen star, said the hermit, and thou shalt only light on some foul jelly, which, in shooting through the horizon, has assumed for a moment an appearance of splendour."

In 1846, *Scientific American* carried a report of an apparently doubly spectacular meteor fall. "It appeared larger than the Sun, illuminated the hemisphere as light as day," said the account. And when it fell "a large company of the citizens immediately repaired to the spot and found a body of fetid jelly, four feet in diameter." Even *Nature* (1) has published a paper on the mysterious gunge, not always associated with falling stars. The author, T. McKenny Hughes, described *pwdre ser* as "a mass of white translucent jelly lying on the turf."

The term *pwdre ser* comes from Wales, bestowed by Welsh shepherds who made many observations on the "rot of the stars." They also described the disappearance of the jelly, by evaporation or some other process, within a few hours of its arrival. *Pwdre ser* was most often seen in the early morning, encouraging the folk belief that it had probably been projected onto Earth by a shooting star unobserved during the hours of darkness.

However, as Hughes pointed out long ago, the association between astronomical events and the discovery of perplexing gels may be purely circumstantial. If people are searching intently for fragments of fallen meteor, they will be more than usually vigilant about any unfamiliar goo on the ground and will tend to give it special significance. This is even more likely to happen to people already conscious of the mythology of *pwdre ser.*

So what exactly is this stuff? Its presence has been reported sufficiently often, and sufficiently reliably, to exclude explanations based purely on delusion or optical illusion. Nieves-Rivera and White argue that there is probably more than one cause. While bird vomit and other nonmicrobiological substances may have contributed to the folklore, other sightings are probably attributable to organisms such as the cyanobacterium *Nostoc.* Members of the Tremellaceae are prime candidates, too—not least *Tremella lutescens* Fr., sometimes known as fairy butter, which is often found in a yellow, gelatinous form after heavy rain. *Tremella concrescens* also forms pale translucent globs in the axes of tufts of grass.

"Myxomycetes too are known by their close encounters with humans," the two mycologists wrote. "In 1973, residents of a small suburb in Dallas, Texas, experienced terror at the sight of moving bright yellow plasmodia of a myxomycete, *Fuligo septica* (L.) Wigg. This motile mass of protoplasm was immediately mistaken as an alien entity in the form of microbes that was starting an invasion of the Earth. The news kept many US citizens spellbound and encircled the nation, similar to Orson Welles's classical radio transmission of an alien invasion on Halloween's Eve in 1938."

Elsewhere, other myxomycetes have attracted human curiosity because of their seemingly extraterrestrial origin. Locals in the state of Veracruz in Mexico refer to two of them, *Enteridium lycoperdon* (Bull.) Farr and *Enteridium septica,* as caca de luna (moon's excrement). However, this epithet does not deter them from frying and eating the myxomycetes' immature fruiting bodies.

One of the most recent sightings of *pwdre ser* was in Scotland on 29 March 2004 by geologist Bill Baird. Climbing with his friend Ronnie Leask, he made the observation as they were approaching the 747-m summit of Meall Mor, which lies about 2 km north of the western end of

Loch Katrine. The wind was dry and breezy with a fairly cloudy sky and a temperature a few degrees above freezing. There were drifts of snow from a previous snowfall.

"As they both took a breather before going on to the summit, Ronnie poked one of several pieces of snow lying on the grass and remarked that it was the first they had come to," says an account in the Autumn 2004 issue of *The Edinburgh Geologist*. "Bill noticed that the reaction of the snow fragment to a poke with Ronnie's stick was unusual. He bent down and picked the piece of 'snow' up and found that it had the consistency of a firm blancmange. The appearance even at arm's length was still absolutely the same as that of a piece of settled fragmented snow bank but the tactile message was of something entirely different."

What they had apparently found was a sample of star jelly. Certainly the material matched the description given by Hughes in his *Nature* paper: "a mass of white translucent jelly lying on the turf." There was, however, a difference in size; Hughes described his pieces as about as large as a man's fist, while those on Meall Mor were bigger than a half brick. Also, Hughes and Leask detected no smell, a common feature in past reports.

The most surprising aspect of the *pwdre ser* story is that, despite a considerable amount of literature (more than ever since the advent of the Web), sightings have received very little scientific attention. The vast majority of reports, whether by mycologists or nonmycologists, have been purely descriptive. Identification, when attempted at all, has been based on purely morphological criteria. There seem to have been no studies at all using modern molecular techniques.

Perhaps microbiologists, being serious-minded people, consider the entire subject simply risible—or worse?

References

1. **McKenny Hughes, T.** 1910. Pwdre ser. *Nature* **83**:492.
2. **Nieves-Rivera, Á. M., and D. A. White.** 2006. Ethnomycological notes. II. Meteorites and fungus lore. *Mycologist* **20**:22–25.

8 Botox and Dairy Cows

Compared with many other infections, botulism is considered to be rather well characterized. Textbooks portray the neurotoxins of *Clostridium botulinum*, the pathogenesis and symptomology of the disease, and its investigation as one of the tidiest sectors of medical microbiology.

Indeed, the organism and its toxins are so well understood that purified, diluted neurotoxin A is now widely administered to treat medical conditions ranging from strabismus and torticollis to spasticity and writer's cramp. It is also given to combat wrinkles, frown lines, and other signs of ageing, not only in the rich and famous but also among rapidly increasing numbers of people who are neither rich nor famous but who follow fads and fashions. Botox, the most poisonous of all known biological substances, would surely not be deployed so indiscriminately if we did not have a profound grasp of the toxin(s) and the molecular mechanisms responsible for botulism.

In certain respects, such confidence appears to be amply justified. However, a report a few years ago on a suspected outbreak of the disease among cows on an English dairy farm delivered a severe blow to my faith in these matters. Out of a herd of 164 animals affected during the incident, 141 died, yet the source of the problem was not conclu-

Animalcules: the Activities, Impacts, and Investigators of Microbes

sively traced, nor were its implications for human health fully clarified. The investigators' dedicated efforts to achieve both of these ends were stymied by difficulties ranging from the inadequacy of laboratory tests for botulism to uncertainties about the pharmacokinetics and pharmacodynamics of *C. botulinum* toxins.

The first indication of disease on the farm was when a single cow began to move in an uncoordinated way. Others then started to appear listless and became unsteady on their back legs. Some lay down and showed signs of weakness in their hindquarters when stimulated to rise. Within a few days, 6 animals had died and a further 20 were developing clinical signs. There were other changes indicative of botulism. The affected animals developed photophobia, and the carcasses of dead cattle had unusually prominent eyes.

Laboratory tests revealed that the ailing cows showed both neutrophilia and hyperglycemia, but these changes were not consistently accompanied by any other hematological or biochemical abnormalities. Taken together, and after the elimination of other possible explanations, such as mineral imbalance and spinal meningitis, the clinical signs pointed strongly to a diagnosis of botulism. Details of the epidemiology of the outbreak supported this conclusion.

Unfortunately, despite a strenuous search for a definitive explanation, the investigators were unable to go much further. Serum samples taken from three of the dying cows, screened by the conventional bioassay in mice, failed to show any evidence of *C. botulinum* neurotoxins. Similarly, toxin(s) could not be detected in samples of the silage, effluent from the silage, or the "molasses sweet wheat" that had been fed to the cattle or in water from their water troughs, nor could the bacterium itself be recovered from these samples.

Among possible sources of the (presumed) toxin, dietary ingredients purchased by the farmer were eliminated on the grounds that no other farms appeared to have been affected. The rapid increase in the number of cows that developed the illness indicated that either the drinking water or a dietary component mixed on the farm was to blame. Since there was no evidence of contamination by putrid material in the pasture, the general environment, or the water troughs, it was assumed that silage was to blame.

The investigators, from the Veterinary Laboratories Agency in Lough-borough and other United Kingdom centers, were thus unable to confirm the precise cause of the incident, its source, or the type of the presumed *C. botulinum* toxin. They are not alone in reaching such an impasse.

"A failure to detect the toxin in affected cows or to identify the affected component of the diet is unfortunately typical of this condition," they wrote (1). "Previous reports have discussed the inadequacy of the laboratory tests available for the diagnosis of botulism, and research into improved diagnostic techniques is long overdue."

There are, in fact, enzyme-linked immunosorbent assay methods that can detect antibodies against a range of *C. botulinum* toxins and others that are used to reveal C and D neurotoxins, but they are apparently no more sensitive than the mouse bioassay used in this investigation, which is far from ideal.

In the dairy farm outbreak, the development of signs of illness over a total period of more than 2 weeks suggested that the animals had been exposed to the toxin in their feed or water for no more than a few days, and possibly a single day. Moreover, the disease progressed more rapidly among the high-yielding cows (those giving larger quantities of milk) than among the low-yielding animals. This probably reflects the fact that those with larger appetites received greater doses of the poison.

Other gaps in knowledge made it difficult to evaluate the implications of the outbreak for human health. First, while cattle are usually affected by *C. botulinum* type C or D, they can also succumb to type B, one of the strains that are responsible for the disease in humans. Secondly, the pharmacodynamics of type C and D toxins in humans are unclear. While there have been sparse reports of type C and D infections in humans, the type C toxin seems to be undetectable in most clinical specimens.

The pharmacokinetics of the neurotoxins is surrounded by comparable uncertainty. "The toxins are released as complexes with non-toxic proteins and are activated by protease enzymes," the investigators pointed out. "Toxin is usually undetectable in blood from clinically affected cows, suggesting that the risk of its transfer to milk is low. However, the highest risk of toxin being transferred into milk or meat may be soon after the animal's exposure, when the circulating titre of activated toxin in blood is highest, before the activated toxin has become fixed in motor end plates, before any clinical signs have been observed and before the toxins have had time to degrade naturally."

Moreover, the persistence of activated toxins in dairy produce or meat may depend upon the conditions under which these foods are processed and stored. It will not necessarily be the same as the persistence of the toxins in putrid material. In addition, the oral bioavailability of the toxin complexes secreted by *C. botulinum* is likely to differ from that of the activated toxins.

Given the many unanswered questions surrounding this outbreak, it is hardly surprising that the authorities adopted a precautionary approach. For example, they placed restrictions on the movement and sale of milk from the farm until 14 days after the onset of the last clinical case. The investigators urged that comparable measures be adopted in any similar circumstances in the future.

The lacunae in knowledge revealed by this study are complemented by others. They were highlighted by the investigators of a previous farm outbreak who did detect type C1 neurotoxin in bovine muscle tissue (4). "As there are no pathognomic signs of botulism at post mortem examination, such meat would pass abattoir inspection," they wrote. "Unfortunately, there appears to be little information about the occurrence and consequences of residual botulinum type C toxin in meat from affected animals. The meat from suspect cases of botulism, or from healthy animals which have been exposed to a source of botulinum toxin, should therefore be considered a potential health risk to human beings or animals if it has not been cooked properly."

Meanwhile, medical anxieties about so-called Botox treatment, the fastest growing cosmetic procedure in the United States, have been growing. One report (2) indicated that the toxin can cause not only headache, nausea, and drooping eyelids, but even cardiac arrhythmia and myocardial infarction. It brings cosmetic benefits by blocking the release of acetylcholine, apparently by cleaving a synaptosome-associated protein, SNAP-25. London neurophysiolgist Peter Misra has warned that the toxin's long-term effects remain unknown (3).

There is, of course, no significant link between the uncertainties surrounding botulinum neurotoxin on the farm and the confidence with which antiwrinkle advocates peddle their wares at Botox parties. Is there?

References

1. **Cobb, S. P., R. A. Hogg, D. J. Challoner, M. M. Brett, C. T. Livesey, R. T. Sharpe, and T. O. Jones.** 2002. Suspected botulism in dairy cows and its implications for the safety of human food. *Vet. Rec.* **150:**5–8.

2. **Larkin, M.** 2002. Promotion of cosmetic botulinum toxin A frowned upon. *Lancet* **360:**929.

3. **Misra, V. P.** 2002. The changed image of botulinum toxin. *Br. Med. J.* **325:**1188.

4. **Neill, S. D., M. F. McLoughlin, and S. F. McIlroy.** 1989. Type C botulism in cattle being fed ensiled poultry litter. *Vet. Rec.* **124:**558–560.

9 Fiction, Fact, and Reality

Journalists working in radio insist that the best pictures come from their medium, rather than television. Images conjured up by words can be far more vivid than those on a TV screen. Here, I want to suggest that the realities of infectious disease are sometimes portrayed most graphically, not in textbooks, but in novels and in factual works whose authors have moved away from a strictly scientific mode of writing.

Consider the following:

Rieux found his patient leaning over the edge of the bed, one hand pressed to his belly and the other to his neck, vomiting pinkish bile into a slop-pail. After retching for some moments, the man lay back again, gasping. His temperature was 103, the ganglions of his neck and limbs were swollen, and two black patches were developing on his thighs. . . . 'It's like fire', he whimpered. 'The bastard's burning me inside.' He could hardly get the words through his fever-crusted lips and he gazed at the doctor with bulging eyes that his headache had suffused with tears.

Those words come from *La Peste (The Plague)* by Albert Camus, published first in French in 1947 and then in English the following year (3). It is a stunning piece of work, whose verisimilitude in showing the effects of *Yersinia pestis* on individual victims is enhanced by the author's menacing portrayal of the impact of the infection on an entire community, the Algerian port of Oran. At a third level, the book describes the epidemic as an allegory for invasion and occupation by the Germans and the defeat of France.

Here is one such passage, illustrating the sense of desolation that grew stronger as the disease became firmly established and the town's gates were closed:

By mid-August, the plague had swallowed up everything and everyone. No longer were there individual destinies; only a collective destiny, made of plague and the emotions shared by all. Strongest of these emotions was a sense of exile and of deprivation, with all the cross-currents of revolt and fear set up by these.

There is an obvious parallel here with *A Journal of the Plague Year* by Daniel Defoe, published in 1722. Though based on actual events in London nearly 60 years previously, and drawing on the memories of people who had lived through the epidemic, the book's disturbing power comes primarily from Defoe's imagination. Both it and *La Peste* should be required reading for anyone researching or dealing with plague today.

David Anne, the author of *Rabid* (1), may not stand alongside Camus or Defoe in the annals of great literature. Nevertheless, there is some gripping writing in his story of John and Paula Denning, who pick up Asp, a lovable stray dog, while on holiday in France and smuggle it back into Britain in defiance of quarantine regulations.

Predictably perhaps, Asp proves to have rabies, and the ensuing events afford many opportunities for the display of blood and spittle. Here, for example, is the scene when, after the dog has been unwell for several days, John enters the kitchen to find a scene of destruction and apparent vandalism:

But no human agency, no mechanical device, had perpetrated this revolting mess. Crouched in a corner, oblivious to John, was Asp. Her coat, matted in filth, stood starkly up about her like so many spikes. Her eyes stared straight into madness with a blazing malignancy. With first this paw, then that, she tried incessantly to wipe away the loathsome glutinous dis-

Animalcules: the Activities, Impacts, and Investigators of Microbes

charge dribbling from her mouth. In vain. It clung about her muzzle with
an obscene stickiness beyond all normal slaver. Not from fright alone but
on account of something far more perversely subtle, John shuddered.

Over the top? Maybe, yet I rarely read about rabies nowadays without these and other images from David Anne's book slipping into the back of my mind. So, too, with Paul de Kruif's description in *Microbe Hunters* (4) of the horror of diphtheria, and the impotence of doctors, before the development of diphtheria antitoxin:

The wards of the hospitals for sick children were melancholy with a for-
lorn wailing; there were gurgling coughs foretelling suffocation; on the
sad rows of narrow beds were white pillows framing small faces blue with
the strangling grip of an unknown hand. Through these rooms walked
doctors trying to conceal their hopelessness with cheerfulness; powerless
they went from cot to cot—trying now and again to give a choking child
its breath by pushing a tube into its membrane-choked windpipe.

This is the reality that finds no place in the sober prose of the typical textbook. Indeed, for all its influence in attracting countless microbiologists into their trade, *Microbe Hunters* has itself been criticized as a piece of popular science for being unduly emotive.

Some authors, however, have concluded that imagery can be both legitimate and effective, even in a textbook. Here is William Boyd, writing about viral pneumonia in *Pathology for the Physician* (2):

The lungs were voluminous and covered in a fibrinous exudate. They had a
firm or rubbery feel, and areas of consolidation could be felt in every lobe.
Excised portions when placed in water floated a short distance below the
surface, but did not sink to the bottom as in lobar pneumonia. Not only
were they red and congested, but numerous minute hemorrhages were
scattered over the surface. The cut surface was a vivid red, with here and
there splashes of a darker colour. The lung was wet and waterlogged. . . .

This is a more colorful piece of descriptive writing than one usually finds in a textbook, yet the additional quality does not detract from the credibility or precision of the text any more than objective science impairs the appeal of creative writing. There is much to be learned about scientific method, for example, from books such as Sinclair Lewis's *Arrowsmith* (5) and John Rowan Wilson's *The Double Blind* (6).

My own favorite example of a piece of text with great power comes from neither a novel nor a textbook but from the autobiography of science fiction pioneer H. G. Wells. Here is part of his account of the hemoptysis that he experienced as a young man many years before:

Every time I coughed and particularly if I had a bout of coughing, there was the dread of tasting the peculiar tang of blood. I can remember as though it happened only last night the little tickle and trickle of blood in the lungs that preceded a real hemorrhage. There was always the question of how big the flow was to be, how long it would go on, and what was going to be the end of it this time. As one lay exhausted, dreading even to breathe, there was still the doubt whether it was really over.

Vivid writing? I believe so, although those words have a particularly keen impact on me because, at the age of 18, I suffered a very similar experience. Fact or fiction, reading is always a subjective process.

References

1. **Anne, D.** 1977. *Rabid.* W. H. Allen, London, United Kingdom.

2. **Boyd, W.** 1958. *Pathology for the Physician.* Lea & Febiger, Chicago, IL.

3. **Camus, A.** 1948. *The Plague.* Hamish Hamilton, London, United Kingdom.

4. **de Kruif, P.** 1926. *Microbe Hunters.* Harcourt Brace, New York, NY.

5. **Lewis, S.** 1925. *Arrowsmith.* Grosset & Dunlap, New York, NY.

6. **Rowan Wilson, J.** 1960. *The Double Blind.* Collins, London, United Kingdom.

10 Microbiology for Gastronomes

Antiquarian book collector Guy Meynell was bothered about foxing, those yellow-brown marks often found on the pages of elderly books, so he asked a few dealers what caused these unsightly (and irremovable) stains. "Oh," they all said. "That's recognized in the trade as foxing." "I know what it's called," Meynell replied, "but what actually is it?" No one knew.

Guy Meynell was not only a bibliophile, but also a microbiologist, so he got busy, studying sheets of paper taken from books published between 1842 and 1919. Using both low-power light microscopy and higher-power fluorescence microscopy, he soon discovered that foxing was a result of fungal infection. "Foxed areas invariably showed fungal hyphae weaving around, but not within, individual cellulose fibres, whereas the surrounding normal paper was almost free of mycelium," he reported in *Nature* (1). Problem solved.

I have had a similar difficulty with corking. From time to time, when opening a bottle of wine, one encounters a characteristic moldy flavor and odor. The wine is said to be "corked" and has to be poured down the sink. But what does this term mean?

"Having an unpleasant smell and taste as a result of being kept in a bottle sealed with a faulty cork," says my dictionary, but faulty in what respect? What specifically has happened to, or between, the wine and the cork? I have asked several wine waiters and sellers over the years, and all have been as unhelpfully tautologous as Guy Meynell's book dealers: "It's caused by corking."

I know the real answer as a result of attending the 10th European Congress on Biotechnology, held in 2001 in Madrid, Spain. There researchers from two Spanish universities, in Valencia and Badajoz, described their attack on a problem which, despite its considerable economic importance, has until now been little understood.

Their major finding was that cork (isolated from the cork oak, *Quercus suber*) is home to a remarkably rich and complex microbial ecosystem. The Spanish team found no less than 13 different filamentous fungi, 9 yeasts, 13 actinomycetes, and 20 nonfilamentous bacteria in samples of cork. Moreover, three-quarters of the filamentous fungi could synthesize the substance strongly suspected of being the principal cause of corking: 2,4,6-trichloroanisol.

Further studies showed how one isolate, a *Trichoderma* strain, produced the unwanted metabolite from 2,4,6-trichlorophenol. The relevant enzymes were being characterized, and there was now hope of understanding why corking occurs on some occasions but not others and of evolving a strategy to thwart the process.

Iberian food was a strong theme of the 2001 meeting, and at least one presentation highlighted a hitherto unexplored link between gastronomy and health. Consider cecina, a beef product characteristic of León in northwest Spain. Microorganisms play a crucial role in the curing and ripening of cecina over a period of about a year. Indeed, the successive appearance of various micrococci and filamentous fungi, principally of the genus *Penicillium*, indicates that the maturation is proceeding well.

Although the preparation of cecina was originally an entirely natural process, starter cultures are now used. The twin aims are to introduce organisms that contribute to flavor and taste and at the same time inhibit the proliferation of mycotoxin producers. A potential drawback has come to light, however, as medical microbiologists have grown increasingly concerned about the problem of antibiotic resistance. Most widely used starter cultures contain *Penicillium nalgiovense*. This not only enhances the gastronomic appeal of cecina, but also bestows its distinctive white color. Unfortunately, many strains generate penicillin, as well, prompting anxieties that this could contribute to the development of beta-lactam resistance in the gut flora.

As reported in Madrid, researchers at the Institute of Biotechnology in León developed strains of *P. nalgiovense* that do not produce penicillin. They achieved this by both mutation and selection and by disrupting the gene coding for the first enzyme in the biosynthetic pathway. Starter cultures containing the new strains no longer generate the antibiotic, therefore, but do retain the desirable qualities of the wild types. Spaniards can continue to enjoy their cecina without hazard.

For gastronomes everywhere, bitterness is an ambiguous quality, sometimes enhancing and sometimes impoverishing the quality of foods and drinks. Lemon and other citrus juices are a prime example, their distinctive appeal being seriously impaired by the unpleasant bitterness attributable to limonoids, especially limonin. This lowers the quality and thus the value of the juices. Moreover, bitterness that develops after extraction of the juice means that some varieties of oranges, lemons, and grapefruit are used far less widely than would otherwise be possible. This "delayed bitterness" is caused by conversion of the nonbitter precursor, limonate A-ring lactone, to limonin.

The economic consequences for Spain and Portugal have been considerable, and a major research effort has been under way to combat the problem. One team, at the University of Burgos, has mounted an extensive screening program based on the natural flora of orange peel, citrus compost, and soil. It has yielded several organisms that are likely to be harnessed to remove limonin from citrus juices. The two prime candidates are strains of *Microbacterium flavescens* and *Enterobacter cloacae*, which the research team described as very efficient biocatalysts for debittering juices.

A freshwater microalga, *Haematococcus pluvialis*, is the workhorse that researchers at the University of Santiago have harnessed to make astaxanthin. The alga has long been recognized as a rich source of this ketocarotenoid, which is used as a food additive for the pigmentation of farmed fish and shellfish. Nevertheless, the development of an efficient industrial process based on the organism has proved extremely difficult.

Astaxanthin is produced when green motile cells of *H. pluvialis* are transformed into red aplanospores. However, past efforts to make the pigment in bulk have been bedeviled by very poor growth rates and final cell densities of the organism in conventional culture media, as well as by a lack of understanding of the maximal conditions for astaxanthin production.

Jaime Fabregas and coworkers in Santiago overcame these obstacles and evolved a two-stage, semicontinuous process. In the first phase (developed through conventional optimization of the medium constituents, light, and aeration), green vegetative cells of *H. pluvialis* are grown to considerably higher densities than were achievable hitherto. This is followed by a second phase, designed to engender stress conditions, when the light level and aeration are increased. As nitrogen becomes depleted, the vegetative cells rapidly turn into aplanospores. As much as 5% of their dry weight consists of the pigment.

The Santiago workers moved from a laboratory scale system, in which they established the various parameters, to large-scale (200-liter) flat-panel photobioreactors. This system achieves cell densities and astaxanthin concentrations equal to those obtained on the bench.

Finally, wine. Geographically, climatically, and gastronomically, we know roughly what distinguishes a fine wine from a mediocre one. However, despite the elucidation of the *Saccharomyces cerevisiae* genome in 1997, we know little about the genetic differences between wine yeast strains. We do not fully understand why, as a group, they ferment musts, for example, whereas other industrial and laboratory strains fail to do so.

José Pérez-Ortin of the University of Valencia set out to change all that. He and his colleagues performed DNA array analyses of wine and laboratory yeasts and found (even under conventional, identical growth conditions) that the expression patterns of over 40 genes differed significantly. Their studies of specific genes in yeasts from different wine regions and vintages have revealed much about the history of natural adaptation to the winery environment. Both oenologists and gastronomes should be the beneficiaries.

When I first heard about this work, I was so delighted that I simply had to open a bottle to celebrate. It was not corked.

Reference

1. **Meynell, G. G., and R. J. Newsam.** 1978. Foxing, a fungal infection of paper. *Nature* **274:**466–468.

11 The Double Life
of *Escherichia coli*

In the 1960s (and, in some cases, considerably later), *Escherichia coli* appeared in specialist texts as essentially benign and in popular books as essentially benevolent. A symbiotic occupant of our intestines, this agreeable organism was also a tireless workhorse of science and (through procedures such as the presumptive coliform count in water testing) a guardian of public health. Strains with harmful potential were so rare that they were often scarcely mentioned at all.

Today, that picture seems to have been almost totally reversed. In addition to enteroaggregative *E. coli*, recognized as a serious agent of enteritis in children, and with worldwide headlines for the horrendous serotype O157:H7, which causes both hemorrhagic colitis and hemolytic-uremic syndrome, four groups are now known to be responsible for diarrhea and death. Though the nomenclature still needs to be fully resolved, these groups are classified as enterotoxigenic *E. coli*, enteroinvasive *E. coli*, enteropathogenic *E. coli* (EPEC), enterohemorrhagic *E. coli* (EHEC), and verocytotoxin (Shiga toxin)-producing *E. coli* (VTEC) varieties.

Underlined by Brett Finlay and his colleagues' devilish discovery that EPEC even makes its own receptor, which it then injects into gut epithelial cells before docking (8), the contrast with our previous assessment of *E. coli* is startling. So, was this primordial coliform as innocuous as it appeared to be just 3 decades ago? Has it developed its formidable range of dangerous capacities in the intervening years (as with serogroup O157, which appears to have acquired its toxin gene from *Shigella dysenteriae*), or were medical microbiologists slow to discern its true character?

Part of the answer is that, while most authors writing in the first decades after World War II did acknowledge the role of *E. coli* in urinary tract infections and its frequent presence in pus, they were slow to grasp the organism's gastrointestinal significance. It was 1939 when John Bray began work at the Hillingdon Hospital, Hillingdon, Middlesex, United Kingdom, on infantile enteritis, especially outbreaks of so-called summer diarrhea, which was often fatal. His findings, published in 1945, clearly established that these cases were often associated with *Bacterium coli*, now called *E. coli*.

Shortly afterward, two serotypes were incriminated as particularly likely to cause infantile diarrhea. They were O55 and O111, the latter isolated from frozen feces from nosocomial outbreaks during 1947 in New York State. Efforts to confirm the pathogenicity of these organisms in laboratory animals failed, however, and this precipitated the morally dubious experiment of inoculating *E. coli* O111 into a 2-month-old infant with multiple congenital defects (3). The infant developed severe diarrhea.

It soon emerged that such strains were pathogenic for adults, as well as children, yet their true depredations took a perplexing amount of time to receive appropriate attention in textbooks. Even during the 1950s, when veterinary microbiologists were confirming the importance of *E. coli* as an agent of diarrhea in calves, its clinical significance was often relegated to footnotes.

Especially curious have been those accounts, published in recent decades, that have documented the burden of morbidity and mortality attributable to *E. coli* in developing countries yet have not fully reflected similar grounds for concern in developed nations. It was in 1998 when one authority (1) pointed out that the emergence of groups such as EHEC and VTEC has finally "altered this complacency."

Likewise, the invasiveness of this once-loved animalcule as a cause of meningitis and septicemia in both adults and children took many years to be fully appreciated. As illustrated by several chapters in Max Sussman's monumental *Escherichia coli: Mechanisms of Virulence* (10), it is an alarming paradox that this so-called commensal has now become a model for investigating bacterial virulence.

Few definitive answers are yet available to questions about the evolution of *E. coli* over the decades and its possible acquisition of new pathogenic habits. However, as Sussman also showed, there is now a wealth of molecular genetics that both illuminates present dangers (for example, the linkage of genes for virulence and antibiotic resistance) and promises to clarify what happened during the organism's past.

There is certainly evidence that at least some of the properties now causing deep concern were present before they were first described in the literature. Later data have indicated that human VTEC infections have increased over the past decade in various parts of the world. This prompted one team to reinvestigate *E. coli* isolates taken from calves in Germany in 1965, more than a decade before VTEC strains first came to light and were recognized as agents of human disease. They found that such strains were indeed already present in cattle over 30 years ago (4). However, there have been conflicting claims as to whether the frequency of VTEC in cattle has increased during past decades. It remains unclear whether improved surveillance and detection methods can account to some degree for the reported rise in VTEC infections.

Writing about *E. coli* in *The Chemical Activities of Bacteria* (7), the Cambridge pioneer Ernest Gale said that "this organism is easily grown in large quantities, is non-pathogenic, has wide chemical activities, and has consequently been subjected to more intense biochemical investigation than any other bacterium." One of the benefits of that popularity, vastly extended in the era of molecular genetics, is the existence of countless strain collections. Systematically investigated, these should soon provide insights into many aspects of the evolution of pathogenicity and virulence in *E. coli*.

I even made a modest contribution to one of those repositories myself. It consisted of fortnightly stool samples, handed over to University of Durham researchers Alan Emslie-Smith and Marjorie Wilkinson in the early 1960s, when they were monitoring the diversity of *E. coli* strains and their temporal patterns in a group of eager volunteers. They phage typed and colicine typed the isolates in an effort to determine the significance of characteristics such as fimbriation and adhesion in transient and "dynastic" strains.

That work was an offshoot of the discovery of fimbriae by John Duguid's team at the University of Edinburgh some years previously. Their breakthrough, in turn, soon created a large agenda for the study of fimbriae and the projections later known as pili and their roles in processes ranging from gene transfer to attachment to epithelial surfaces.

The sequencing in 1997 of *E. coli*, the seventh genome to be laid before us, took the study of these and other features into a new dimension. Commentators responded by observing how, much more quickly than with any other organism, we can combine that genetic blueprint with our existing storehouse of phenotypic information to assemble an exhaustive portrait of the primordial coliform.

Greater understanding of enterotoxigenic *E. coli*, EPEC, EHEC, and VTEC and how they evolved is now flowing from this work. Undue optimism concerning the immediacy of those returns should be tempered, however, by one observation. Ernest Gale described *E. coli* as intensively studied over half a century ago, and it has been investigated increasingly extensively ever since. Nevertheless, of the predicted complement of over 4,000 genes that program this deceptively unfamiliar animalcule, we have little or no idea what nearly 40% of them are doing, for good or ill.

More than a century after Theodor Escherich first identified *B. coli commune*, it still offers a formidable agenda for research and understanding. Meanwhile, *E. coli* O157:H7 continues to impact human affairs in a disturbing variety of ways. In 2006, these ranged from a gastroenteritis outbreak in a care home for the elderly in the West Midlands in Britain (2), the precise source of which could not be identified, to a multistate outbreak in the United States associated with the consumption of fresh spinach (5). More recent incidents were a family outbreak in Italy traced to pork salami (6), several cases associated with a swimming pool near Manchester in the United Kingdom (11), and an outbreak in southern Sweden attributed to fermented sausage (9).

Despite its innocent and indeed benevolent past, *E. coli* can clearly be a wolf in sheep's clothing.

References

1. **Acheson, D. W. K.** 1998. Nomenclature of enterotoxins. *Lancet* **351**:1003.

2. **Afza, M., J. Hawker, H. Thurston, K. Gunn, and J. Orendi.** 2006. An outbreak of *Escherichia coli* O157 gastroenteritis in a care home for the elderly. *Epidemiol. Infect.* **134**:1276–1281.

3. **Baldwin, T. J.** 1998. The 18th C. L. Oakley Lecture. Pathogenicity of enteropathogenic *Escherichia coli*. *J. Med. Microbiol* **47**:283–293.

4. **Beutin, L., and W. Müller.** 1998. Cattle and verotoxigenic *Escherichia coli* (VTEC), an old relationship? *Vet. Rec.* **142**:283–284.

5. **Centers for Disease Control and Prevention.** 2006. Ongoing multistate outbreak of *Escherichia coli* serotype O157:H7 infections associated with consumption of fresh spinach—United States, September 2006. *Morb. Mortal. Wkly. Rep.* **55**:1045–1046.

6. **Conedera, G., E. Mattiazzi, F. Russo, E. Chiesa, I. Scorzato, S. Grandesso, A. Bessegato, A. Fioravanti, and A. Caprioli.** 2007. A family outbreak of *Escherichia coli* O157 haemorrhagic colitis caused by pork meat salami. *Epidemiol. Infect.* **135**:311–314.

7. **Gale, E.** 1947. *The Chemical Activities of Bacteria.* University Tutorial Press, London, United Kingdom.

8. **Kenny, B., R. DeVinney, M. Stein, D. J. Reinscheid, E. A. Frey, and B. B. Finlay.** 1997. Enteropathogenic *E. coli* (EPEC) transfers its receptor for intimate adherence into mammalian cells. *Cell* **91**:511–520.

9. **Sartz, L., B. De Jong, M. Hjertqvist, L. Plym-Forshell, R. Alsterlund, S. Löfdahl, B. Osterman, A. Ståhl, E. Eriksson, H. B. Hansson, and D. Karpman.** 2008. An outbreak of *Escherichia coli* O157:H7 infection in southern Sweden associated with consumption of fermented sausage; aspects of sausage production that increase the risk of contamination. *Epidemiol. Infect.* **136**:370–380.

10. **Sussman, M.** 1997. *Escherichia coli: Mechanisms of Virulence.* Cambridge University Press, Cambridge, United Kingdom.

11. **Verma, W., F. J. Bolton, D. Fiefield, P. Lamb, E. Woloschin, N. Smith, and R. McCann.** 2007. An outbreak of *E. coli* O157 associated with a swimming pool: an unusual vehicle of transmission. *Epidemiol. Infect.* **135**:989–992.

12 Not All Cigars and Caviar

The European Federation of Biotechnology's biennial congresses have a uniquely cosmopolitan flavor, never eclipsed by the welter of papers on fashionable topics and techniques. The 2005 event, the 12th European Congress on Biotechnology (ECB12), held in Copenhagen, was no exception. Despite ECB12's overall theme of "Bringing Genomes to Life," many contributions expressed distinctive aspects of the geography, gastronomy, agriculture, health, or trade of the participating countries.

Insights into the fermentation of Kentucky tobacco to make Toscano cigars in Italy were matched by news of improvements in the ripening of sugar-salted cod roe to produce caviar emulsion in Norway. There were offerings on the use of *Yarrowia lipolytica* to recycle waste cooking oil in Spain, on an alkaline protease with industrial potential produced by *Bacillus pumilus* from a hypersaline lake in Turkey, and on high-gravity fermentation to make low-calorie beer in Denmark. Several papers reflected moves toward the European Union's target of deriving 5.75% of all transportation fuel from biomass by the year 2010.

Many participants, unaware of the role of microorganisms in cigar manufacture, found the talk by Marianna Paulino of British American Tobacco in Naples especially appealing. She described the process in which dark fire-cured Kentucky tobacco, grown in Italy, is prepared to make Toscano cigars. The leaves are dipped in water and placed in piles, which then undergo fermentation for 1 to 2 weeks, triggered by their own natural flora. The tobacco is turned over occasionally to dissipate the considerable heat generated and to ensure the homogeneity of the product.

Apart from the release of ammonia, virtually nothing had been known about the fermentation or the microbial community responsible. Paulino and her colleagues have now defined three successive phases in the process. Initially, at low pH, *Debaryomyces hansenii* and other yeasts proliferate and suppress the growth of bacteria. Then, the yeasts disappear as bacteria, especially *Corynebacterium ammoniagenes*, grow exponentially, probably using autolysing yeast cells, as well as residual nutrients from the tobacco. During the final phase, when the temperature rises as high as 70°C, only sporing *Bacillus* species are detectable.

After initially characterizing the flora by morphology and other phenotypic traits, the Naples group applied rRNA gene sequence analysis to screen the isolated organisms. They have since used this information to identify ingredients important for the distinctive flavor of Toscano cigars, as well as to investigate, and perhaps eliminate, potentially toxic components.

Microbiologists who enjoy lamb for their meal before a cigar were interested in a quite different contribution to ECB12. It came from Silas Villas-Boas and colleagues at AgResearch Ltd. in Palmerston North, New Zealand. They have sequenced the genome of *Clostridium proteoclasticum*, their reasons for doing so reflecting both their country's investment in genomics relevant to agriculture and the fact that ruminants constitute the vast majority of farm animals in New Zealand. A member of the *Butyrivibrio-Pseudobutyrivibrio* community of rumen bacteria, *C. proteoclasticum* is thought to play an important role in degrading plant hemicellulose-lignin complexes that limit the digestion of fiber within the rumen.

Sequencing this organism allowed Villas-Boas and his collaborators to identify an array of candidate genes with various activities relevant to the breakdown of fiber. "We have established a footprinting approach for microbial metabolome analysis focused mainly on metabolic intermediates of polysaccharide degradation to provide quantitative information on end products of fiber-degrading enzymes," they said. "We believe our data will complement proteomic analysis of these enzymes

Animalcules: the Activities, Impacts, and Investigators of Microbes

and microarray analysis of gene expression from a series of mutants by providing direct evidence of the metabolic function of key genes involved in fiber degradation."

An ECB12 contribution that reflected a health concern in its country of origin came from researchers at the Universidad Autónoma de Nuevo León in Mexico. They developed and patented a technique for the serological diagnosis of invasive amebiasis that preserves the antigenicity of extracts with high protease content without using enzyme inhibitors. Unfortunately, existing indirect hemagglutinin assays and other methods do not yield consistent results in zones of endemicity. Based on Western blotting, the new technique was highly accurate and did not show cross-reactions in patients with multiple intestinal parasites, such as *Giardia lamblia*.

Two papers were of particular interest to food scientists. In the first, Hanne Vang Hendriksen from Novozymes A/S in Bagsvaerd, Denmark, highlighted a problem that first came to light 3 years ago when 50 to 4,000 ppb of acrylamide was found in French fries and other potato- and grain-based foods prepared at high temperatures. Subsequent studies showed that the precursors of this potential carcinogen were asparagine and reducing sugars.

The Danish group found that asparaginase is an effective means, superior to other possible enzymatic treatments, of lowering the level of acrylamide in several different laboratory models of the foods in question. Obtained from *Aspergillus oryzae* and incorporated in French fries and crisp bread, the enzyme diminishes the acrylamide content satisfactorily without impairing the foods' sensory qualities.

However, there were also indications of a previously unsuspected problem for food technology. Anne Gravesen and others at the Royal Veterinary and Agricultural University of Denmark, Copenhagen, reported that sublethal doses of alcohol induced the attachment of *Listeria monocytogenes* to surfaces at low temperatures. Since alcohol is commonly used as a disinfectant in the food industry, and since such levels may occur in places inaccessible to cleaning, their finding provoked considerable concern.

The attachment of *L. monocytogenes* is thought to be accompanied by enhanced production of exopolysaccharide (EPS). So Gravesen et al. used bioinformatics to search for homologues in this organism of known

EPS genes from other, gram-positive bacteria. Their foray revealed several candidates, such as genes encoding glycosyltransferases. Understanding EPS and biofilm formation should, they believe, help them to devise a strategy to minimize a possibly serious risk to the consumer.

The human gut flora has probably been studied more closely than that of any other site in the entire biosphere, yet ECB12 participants learned of a hitherto-unknown member of this community. Discovered by Kim Holmstrøm and colleagues at Bioneer A/S in Hørsholm, Denmark, it represented a new species and a new genus and was named *Subdoligranulum variabile*. It was strictly anaerobic, nonmotile, nonsporing, and gram negative and had unusual pleiomorphic morphology when observed as colonies on an agar plate. First isolated from a woman patient's feces, it is now known to comprise around 1.6% of the total number of fecal bacteria.

Finally, there was news of a novel use for a very familiar member of the gut flora to make a mussel protein that could find commercial applications as an adhesive under aqueous conditions and in medicine. It was reported by Hyung Joon Cha and a team from Pohang University of Science and Technology in South Korea. They took a protein (type 3, variant A0) from the foot of *Mytilus galloprovincialis* and fused it with a hexahistidine affinity ligand in *Escherichia coli*. The adhesive forces and adsorptive capacity of the recombinant protein were similar to those of similar agents developed experimentally in the past, but it was superior in several respects, including its yield on purification. The Korean researchers believe their protein may prove valuable, not only as a medical adhesive, but also for use in a variety of underwater environments.

Amidst all this internationality, only one group allowed national pride to go too far. "This study reports the first isolation of bacteria from deep-sea sediments," announced a team from New Zealand. They meant "deep-sea sediments in New Zealand."

13 Microbial Versatility in Berlin

My prize for the most unexpected use of a microorganism ever reported at a scientific meeting goes to Henryk Struszczyk and his colleagues at the Institute of Chemical Fibers in Lodz, Poland. As described at the World Congress of Biotechnology held in Berlin, Germany, in 2000, they had contrived to harness bacteria in efforts to improve the quality of sound reproduction.

Admittedly, this was not the original purpose of their work. What they set out to do was to engineer *Acetobacter xylinum* to synthesize modified forms of bacterial cellulose. Having accomplished this goal, they then experimented further with the cellulose produced and found that they could spin its fibers and fashion the material into various forms, including sheets. The absorbance and other attributes of the new material raised the prospect of several medical applications, including wound dressings.

However, conversations with physical scientists specializing in sound reproduction prompted Struszczyk and his team to investigate the electroacoustic properties of membranes composed of the novel bacterial cellulose. The results indicated that the membranes satisfied the relevant requirements and led to the group fashioning their material into loudspeaker cones. The results were excellent.

Such curiosities aside, two of the congress's other major talking points were the engineering of staphylococcal surfaces for biotechnology purposes and the practical exploitation of other bacteria even when they cannot be cultured in the laboratory.

Staphylococcus xylosus and *Staphylococcus carnosus* are two organisms that Stefan Stahl and colleagues at the Royal Institute of Technology in Stockholm, Sweden, with collaborators in Saint-Julien en Genevois, France, had modified for both medical and environmental purposes. They first found that both of these staphylococci were nonpathogenic and could be administered safely in very large doses by both mucosal and subcutaneous routes. Later, they demonstrated good immune responses in mice to recombinant strains carrying, for example, surface-exposed streptococcal antigens.

Stahl and his coworkers then greatly improved their system by targeting the putative vaccine vehicles to the mucosa through codisplay of proteins that bind to structures on the mucosal epithelium. They used an *S. carnosus* strain codisplaying a subfragment of *Vibrio cholerae* toxin B to deliver peptides from the G glycoprotein of (human) respiratory syncytial virus. The recombinant organism elicited excellent antibody responses, protective against later respiratory syncytial virus challenge. This was apparently the first report of such immunization afforded by a food grade bacterium against a pathogenic virus.

In parallel, the Stockholm researchers also explored the possibility that modified staphylococci might prove superior to nonengineered bacteria as bioabsorbents to remove toxic metals from wastewaters. Again, they harnessed *S. xylosus* and *S. carnosus*, this time with surface-exposed chimeric proteins containing polyhistidyl peptides designed for attachment to divalent metal ions. The cells did indeed bind cadmium and nickel ions, suggesting that they could prove valuable both in the bioremediation of heavy metals and as components of biosensors for environmental monitoring.

A third potential application of recombinant staphylococci was as inexpensive diagnostic tools. Here, their surfaces would be modified to express antibody fragments in a functional form. Stahl and his colleagues selected so-called affibodies against targets as diverse as human insulin, *Taq* DNA polymerase, and a human apolipoprotein variant. They then worked on the idea of improving bacterial vaccine delivery systems by the selection and display of protein domains targeting the delivery vehicle to immunoreactive sites.

Berlin participants heard about several organisms, identified by molecular techniques, such as fluorescence in situ hybridization (FISH), that have practical utility even though they resist traditional enrichment and isolation efforts. Holger Daims of the Technical University of Munich in Germany described novel yet uncultivated planctomycetes that convert ammonia to gaseous nitrogen in the absence of oxygen in anaerobic trickling filters. Other groups reported the roles of uncultured bacteria as the dominant nitrite oxidizers in many wastewater treatment plants and in enhancing biological phosphate removal.

Takayasu Tsuchida from Ajinomoto Inc. in Tsukuba, Japan, described a powerful new technique using gel microdroplets and flow cytometry to isolate bacteria that cannot be grown at the bench. After successfully applying it to separate an organism that thrives in Difco broth *(Bacillus subtilis)* from an uncultivable one *(Leuconostoc mesenteroides)*, he used the same approach to isolate unculturable bacteria from activated sludge from a food factory. Phylogenetic analysis based on 16S rRNA gene sequences showed that they belonged to new taxa.

While the importance of many hitherto-unrecognized organisms in processes such as sewage disposal is evident, lack of understanding of their biology has hampered efforts to optimize their performance. Holger Daims and his coworkers in Munich and at Aalborg University in Denmark have combined cultivation-independent methods to that end. They identified target populations based on the sequences of rRNA and functional genes and determined their abundances in bioreactors by semiautomatic quantification with FISH, confocal laser scanning microscopy, and digital image analysis. The morphologies and architectures of the organisms can be revealed by three-dimensional visualization, and their physiological traits can be determined by combined FISH and microautoradiography.

One environmental topic of increasing scientific and public discussion is pollution by substances with undesirable endocrine actions. These include natural hormones, but also industrial chemicals, such as nonylphenolic compounds and bisphenol A. Some are suspected of causing hermaphroditism and other abnormalities in aquatic creatures and, indeed, of having adverse effects on humans, too.

Such endocrine disrupters are not fully degraded in conventional sewage treatment plants. This led Silke Schiewer and others at Aachen University of Technology in Aachen, Germany, to seek to remove them by submerging custom-built microfiltration membranes in the treatment unit. One approach depends upon the presentation of appropriate antibodies, procured from phage display libraries in *Escherichia coli*, on algal surfaces as "antibody biofilters."

There were several reports during the Berlin meeting of successful bioremediation projects in grossly polluted regions of eastern European countries. According to Adam Pawelczyk and coworkers, of Wrocław University of Technology, soil and water contamination in areas once used by Soviet troops was "the most spectacular problem in environmental protection in Poland since 1990." At one abandoned airfield in Szprotawa, they isolated, grew, and then reinoculated indigenous organisms, while optimizing the pH and other conditions, to combat gross pollution with aircraft fuel. Contamination declined to an acceptable level in 3 months.

It was no surprise at a microbiological congress in Germany to hear talk of the delights of beer, and in this case, the good news came from researchers in Berlin itself. A team at NovaBiotec came up with an answer to the beta-glucans (from malt) that clog the filters used to clarify beer and thus add to its cost. German laws going back to 1516 forbid the addition of anything to real beer, so the obvious solution of adding beta-glucanase was verboten.

The NovaBiotec workers tackled the problem by covalently coupling highly purified beta-glucanase to an inorganic carrier in a flowthrough reactor integrated into the brewing process. The stratagem worked and could improve the consistency not only of German beer, but also of beers and wines wherever makers choose to use it.

14 Whither Psychoneuroimmunology?

I am disappointed. Several years after the appearance of an impressive clutch of papers linking vulnerability to infectious disease with state of mind, progress in deepening and harnessing that knowledge has been less impressive. Though psychoneuroimmunology has since become an established subdiscipline of microbiology, with its own journals and worldwide conferences, it has not matured as lustily as seemed likely 3 decades ago.

I vividly recall the occasion in the mid-1980s when friends at the Philadelphia-based Institute for Scientific Information showed me what was happening in one region of their "Atlas of Science," a mapping project that demonstrated the relationships between different sectors of research through the analysis of cocitation patterns. When scientist A cites together papers by scientist B and scientist C, this cocitation establishes an intellectual link between those publications. If this happens increasingly frequently, it can indicate the forging of new conceptual or practical links in the advancement of science.

That was what the Institute for Scientific Information scientometricians had spotted, using techniques described by Henry Small and Eugene Garfield in the *Journal of Information Science* (3), in the nexus between the formerly distinct sciences of psychology and immunology. Not only were practitioners of the former beginning to cite papers in the latter domain, and vice versa, but papers in the two fields were increasingly being cited together.

One of the key publications that triggered this process was written by Richard Totman, Sylvia Reed, and Wallace Craig of the then Common Cold Unit, operated near Salisbury, England, by the Medical Research Council. Published in the *Journal of Psychosomatic Research* (4), it focused on "cognitive dissonance," a fey expression for the uncomfortable regret we often experience after making a difficult decision, which in turn we try to justify in one way or another. By subtly maneuvering experimental subjects into this state of mind, Reed and her colleagues confirmed what had long been suspected but never demonstrated indubitably: that infectious diseases can be modified by psychological factors.

The guinea pigs in the study were 48 Common Cold Unit volunteers. When the project began, people had already been going to the unit for 3 decades, having rhinoviruses dropped up their noses, their antibody levels measured, and the development of colds monitored and statistically scrutinized. Some received potential antiviral drugs, others innocuous placebos. All of this work was based on the idea that immune status and the virulence of the invader were what determined the outcome of rhinovirus infection. The brain played no part in the drama.

However, it was the minds of their 4 dozen subjects that interested Reed and her collaborators. They argued that a difficult decision to commit oneself to an unpleasant medical treatment might, via cognitive dissonance, activate the (probably subconscious) search for justification. Optimal justification would, of course, mean that the drug or procedure proved successful: "I'm glad I went for treatment, after all."

So the investigators hatched a plot. They inoculated all 48 volunteers with two rhinoviruses and told them they would probably develop moderate colds as a result. They then asked 23 of these individuals whether they would like to receive an antiviral drug (actually a placebo) that was on trial.

To make the decision more difficult, the researchers also deceived these volunteers into believing that, after treatment, they must have their stomach juices investigated by gastric intubation. (Pilot studies had helped in formulating a choice that would result in about half the subjects opting to take the "drug" and about half declining it.) The

Animalcules: the Activities, Impacts, and Investigators of Microbes

remaining 25 were controls who received neither drug nor the stomach tube threat but were given no choice.

According to cognitive-dissonance theory, the researchers reasoned, those confronted with the decision would show dissonance, whether or not they were given the "drug," and the others would not. Thus, self-justification might mean that those faced by the hard choice would manifest less evidence of common-cold infection than the control group. Clearly, anyone going to the Common Cold Unit at that time experienced some dissonance in justifying his or her 10-day stay, but Reed and her coworkers took account of this and designed a scenario that created a strikingly decisive division to test their hypothesis.

The result was precisely the opposite of that predicted. The psychological condition into which the volunteers had been manipulated did affect the common colds in the two groups, but not in the expected fashion. Corrections were made to allow for bias due to the effect of preexisting antibodies, yet despite this, the symptoms (as independently assessed by a clinician) were significantly more severe in the individuals confronted by choice. There was even a difference, in the same direction, in the amounts of virus shed by the two groups, though this was not considered statistically significant.

Reed and her colleagues concluded that the psychological manipulation of their volunteers did indeed modify the manifestations of the common cold. However, with the original hypothesis demolished, they had to seek another explanation for the observed effects. Ethical undesirability led them to reject a further test which could take them nearer the truth—use of a group given compulsory "treatment" and threat of intubation.

Without the information that might emerge from comparison with this third category, the Salisbury group could only suggest that difficult decision making leads to anxiety, which in turn heightens symptoms. One possibility was that these volunteers feared they were in for particularly severe colds and that this belief really did influence their respiratory reactions.

These findings and speculations occurred 30 years ago and appeared to presage a paradigm shift in our understanding of infectious disease. Commentators confidently asserted that the Common Cold Unit work, together with David Kissen's studies in Glasgow, Scotland, on the effect of stress on tuberculosis, indicated that the mind was probably a potent force acting on all infections. Hitherto, medical microbiologists had ignored a crucial independent variable in framing their mechanistic accounts of infectious disease. Now it was clear that the textbooks were going to have to be rewritten and treatment redirected accordingly.

But what has actually happened since then? There have, of course, been further developments. One particularly striking contribution came from Sheldon Cohen and others in a paper in the *Journal of the American Medical Association* (1). They reported that more diverse social networks were associated with greater resistance to upper respiratory tract illness. Using healthy volunteers, as in Sylvia Reed's work, they demonstrated that those with more types of social ties (such as friends, family, work, and community relationships) were less susceptible to colds triggered by inoculation with rhinoviruses.

We also know considerably more than was apparent in 1997 about communication between the brain and the immune system through cytokines expressed in the brain and neuropeptides expressed in immune cells. Among the first neuropeptides to be found in lymphocytes were corticotrophin and beta-endorphin. Writing in *The Lancet* (2) shortly after their discovery, David Jessop of Bristol University pointed out that "the wide range of immune functions exerted by these neuropeptides, and the presence of all major classes of opioid receptors in immune tissues, provide evidence for the potential complexity and importance of peripheral opioid/immune interactions."

However, despite advances of this sort at the molecular level, we seem not to have attained anything like a full understanding of the phenomena that signaled the emergence of psychoneuroimmunology in the first place. The textbooks have not been rewritten. Were we naïve or imprudent to imagine that they would be?

References

1. **Cohen, S., W. J. Doyle, D. P. Skoner, B. S. Rabin, and J. M. Gwaltney, Jr.** 1997. Social ties and susceptibility to the common cold. *JAMA* **277:**1940–1944.

2. **Jessop, D. S. 1998.** Beta-endorphin in the immune system—mediator of pain and stress? *Lancet* **351:**1828–1829.

3. **Small, H., and E. Garfield.** 1985. The geography of science: disciplinary and national mappings. *J. Info. Sci.* **11**:147–159.

4. **Totman, R., S. E. Reed, and J. W. Craig.** 1977. Cognitive dissonance, stress and virus-induced common colds. *J. Psychosom. Res.* **21**:55–63.

II. The Ecological Context

15 Communal Diversity in Biofilms

Seldom, if ever, has a branch of science moved so swiftly to acknowledge how inadequately it has previously portrayed its subject material. Within barely 2 decades, microbiologists have had to contend with burgeoning evidence that virtually all microbial activity in nature occurs, not in the idealized abstractions of the textbook, but in the highly organized communities known as biofilms. This is a revolution, not evolution, in understanding.

True, the change can be likened to the earlier recognition that pure cultures are rare in nature, which is characterized instead by mixed populations and consortia, but that was a gradual realization, one that extended, but did not radically change, the laboratory study of animalcules. By contrast, the discovery of biofilms and their sophisticated communities represents a true paradigm shift. It is transforming the ways in which microbial activity is investigated, comprehended, and described, as well as providing new perspectives for textbooks now being written.

One sign of the revolution is the increasing frequency with which the term biofilm appears in the titles of papers. Equally telling is the wide variety of subjects dealt with in these reports, which embrace both undesirable and beneficial activities. Topics represented in a recent single volume of the *Journal of Applied Microbiology*, for example, include the development of biofilms in storage cases for contact lenses, their growth in photographic-processing tanks, and their role in inhibiting the corrosion of steel.

The man responsible more than any other for focusing attention on the real lifestyle of microorganisms in the biosphere was William Characklis of Montana State University in Bozeman. I remember him at a Dahlem Conference in Berlin in 1984 trying, not entirely successfully, to persuade other participants of the vast significance of biofilms. Even some of those who were relatively receptive to his ideas wondered whether he had gotten the whole thing out of proportion.

Characklis later established a laboratory to investigate biofilms, an interdisciplinary center reflecting his own robust determination to straddle both microbiology and engineering. Sadly, he died in 1992 at the age of only 50, though the work at Bozeman now continues under the leadership of William Costerton.

In recent years, Costerton and his colleagues have done much to establish that even those textbooks that do contain illustrations of biofilms are naive in presenting them as random assortments of organisms stuck together with mucilage. Using confocal microscopy, the Montana team has demonstrated well-organized microcolonies, often mushroom shaped, enclosed in a polymer matrix. They are separated by water channels that serve as an elementary circulatory system to remove waste and deliver nutrients.

Cross-feeding between different species is just one aspect of the complexity of such communities. In addition, microorganisms in biofilms can express genes that are not manifest in isolated cells, while symbiotic metabolic activities are often greater than those of the individual members. The heterogeneity of biofilms is such that they form anodes and cathodes, which may play a role in corrosion.

David Davies and colleagues at Bozeman and elsewhere helped to reveal the cell-to-cell signaling by which the inhabitants of a *Pseudomonas aeruginosa* biofilm coordinated their behavior (1). They identified a quorum-sensing molecule, an acylated homoserine lactone, as a key signal in the development of the biofilm's elaborate architecture. Mutants unable to produce this lactone could still attach to surfaces and initiate biofilm formation but failed to create intercellular spaces and thus the necessary communal structure.

Three discoveries, of biofilms as undesirables in contact lens cases and photoprocessing tanks and as desirables in preventing metal corrosion, typify the extraordinary range of activities that have come to light over the past decade. Awareness of the first has emerged from the work of Louise McLaughlin-Borlace and colleagues at the Institute of Ophthalmology in London.

The roles of *P. aeruginosa* and *Acanthamoeba* in causing microbial keratitis, a rare but severe complication of contact lens wear, have been known for some years. The puzzle had been why these organisms can be so difficult to eradicate, even with meticulous cleaning and disinfection. The answer, it seems, lies in bacterial biofilms of the type which the London team found in the storage cases of 17 of 20 patients with microbial keratitis (5). In contrast, only 11 of the patients had biofilms on their lenses, where their density was also lower.

Contact lens cases apparently provide an excellent environment for these exquisitely organized mixtures to develop, persist, and serve as a reservoir for the release of pathogens. Like all biofilms, their polysaccharide architecture protects them from antiseptics (in this instance, hydrogen peroxide and chlorine release preparations). The same is true of antibiotics, to which biofilm inhabitants are phenomenally resistant.

Another splendid environment for biofilm life is provided by the sequences of polyvinyl chloride-walled flowing-water tanks used in the photographic processing industry. Especially suitable are the rinsing vessels, which come after those containing developer and fixer and which offer instead a neutral pH and pleasant warmth (30 to 35°C).

By sampling tanks of this sort, Hilary Lappin-Scott and coworkers at the University of Exeter and at Kodak in Harrow, United Kingdom, found a rich biofilm community of bacteria, filamentous fungi, and yeasts (2). This can not only foul, block, and slow down the system, but also damage its products, by detaching from the walls and causing streaks on prints, for example.

Biofilms bring benefits, too. Several groups have reported that, in addition to those which corrode pipes, cooling towers, and ship hulls, others prevent damage of this sort. Thomas Wood and a group at the University of California, Irvine, have characterized the phenomenon in experiments showing that *Pseudomonas fragi* and *Escherichia coli* biofilms can inhibit the corrosion of steel by up to sevenfold (4). Confocal microscopy and other evidence indicated that a certain minimum thickness or density is required for this inhibition to occur. Among other positive contributions, *Pseudomonas fluorescens* biofilms safeguard certain plant roots against invasion by pathogens. Also, much of the cleansing and detoxification in sewage disposal is accomplished, not by free-living organisms, but by biofilm communities. However, many other biofilms, whether on medical implants or underground water pipes, are both unwelcome and inaccessible. What can be done about them?

A variety of tactics have been under development, especially at the Bozeman laboratory. One is to zap the slime-enveloped consortia with low-dose electric currents. A second is to coat catheters with antibiotics to prevent the associations, with their wondrous microarchitecture, from developing in the first place.

Another intriguing technique is being evolved by David Stickler and a group at the Universities of Wales and Birmingham in the United Kingdom: axially propagated ultrasound. They are seeking an alternative to biocides (environmentally unfriendly and comparatively ineffective) and dismantling and mechanical cleaning (tedious and costly) to remove biofilms from cooling-water systems. Tested on *Proteus mirabilis* biofilms in water-filled glass tubes, this approach worked even better than a sonic cleaning bath.

An arguably more subtle alternative would be to interfere with signals of the sort David Davies and his collaborators have discovered in *P. aeruginosa*. This scenario also offers the prospect of encouraging, as well as discouraging, the formation of biofilms. The really smart way of dealing with these ubiquitous communities of animalcules may be by blocking, enhancing, and modifying their communications. Plausibly, one of the drawbacks of biofilms, their tendency to protect organisms from antimicrobial attack, might even be turned to advantage. Following the demonstration that sublethal concentrations of antibiotics promote biofilm formation, coupled with an indication of the molecular mechanisms underlying this response (3), researchers have been working on ways of interfering with that very process. It is an ingenious idea.

References

1. **Davies, D. G., M. R. Parsek, J. P. Pearson, B. H. Iglewski, J. W. Costerton, and E. P. Greenberg.** 1998. The involvement of cell-to-cell signals in the development of a bacterial biofilm. *Science* **280**:295–298.

2. **Elvers, K. T., K. Leeming, C. P. Moore, and H. M. Lappin-Scott.** 1998. Bacterial-fungal biofilms in flowing water photo-processing tanks. *J. Appl. Microbiol.* **84**:607–618.

3. **Hoffman, L. R., D. A. D'Argenio, M. J. MacCoss, Z. Zhang, R. A. Jones, and S. I. Miller.** 2005. Aminoglycoside antibiotics induce bacterial biofilm formation. *Nature* **436**:1171–1175.

4. **Jayaraman, A., A. K. Sun, and T. K. Wood.** 1998. Characterization of axenic *Pseudomonas fragi* and *Escherichia coli* biofilms that inhibit corrosion of SAE 1018 steel. *J. Appl. Microbiol.* **84**:485–492.

5. **McLaughlin-Borlace, L., F. Stapleton, M. Matheson, and J. K. Dart.** 1998. Bacterial biofilm on contact lenses and lens storage cases in wearers with microbial keratitis. *J. Appl. Microbiol.* **84**:827–838.

16 Biofilm Life

While medical microbiologists in particular still love to isolate organisms in pure culture, most of their kindred now recognize the limitations and artificiality of this simplistic craft. The greater reality is, of course, the complex world of polymicrobial communities, such as biofilms. Scrutiny of these heterogeneous populations is now providing both deeper insights into the sophistication of microbial life and pointers toward possible avenues for cooperation and control (1, 2).

Many examples emerged during a Society for Applied Microbiology (SAM) meeting held in Edinburgh, Scotland, in 2006. They were from domains as disparate as the dynamics of the human intestinal flora and the bioremediation of contaminated soil.

Biofilms can be a significant force driving microbial evolution, too, as illustrated by one of the SAM contributions. It came from Soren Molin and coworkers at the BioCentrum DTU in Lyngby, Denmark. They established, as a model consortium, a biofilm consisting of just two organisms. These were *Acinetobacter* sp. strain C6, which was able to mineralize benzyl alcohol, and *Pseudomonas putida* KT2440, which could mineralize benzoate but not benzyl alcohol because it had been cured of the TOL plasmid.

Grown together in a flow cell and supplied with benzyl alcohol as the sole source of energy and available carbon, the bacterial duo established a commensal relationship. *Acinetobacter* oxidized the substrate to benzoate, some of which was excreted (due to a bottleneck in the degradation pathway) and was then used by the pseudomonad as its carbon and energy source.

However, this was not the full story. "Despite the apparent commensality of the association, a niche developed in the structured film, in which the wild-type cells of KT2440 could not persist," Molin reported. "Reproducibly, variants appeared, displaying a number of phenotypic changes, including the capacity to occupy the free niche and exploit efficiently the released benzoate from C6." Genetic analysis confirmed that there were alterations in the composition of the cell surfaces of the variants. Moreover, the selective forces resulting in the emergence of the variants appeared to be restricted to the biofilm. They were present only rarely, if at all, in homogeneous suspended cultures.

In the early years of research on bacterial biofilms, they have sometimes been portrayed as passive occupants of inactive surfaces, whether oil rigs or shellfish in the oceans or tissues or implants in the body. Another SAM contributor, Jeremy Webb of the University of Southampton, argued that this view was seriously incomplete. One of his examples was the subtle interactions between the seaweed *Delisea pulchra* and its epiphytic biofilms. Far from remaining impervious to the presence of the bacteria, the alga partially determines their composition and abundance by releasing furanones that interfere with bacterial signaling.

Webb also described how biofilms undergo intrinsic ontogenic effects, such as regulated differentiation and cell death, which lead much of the structure to disperse and slough away. One such mechanism is the internal production of reactive oxygen or nitrogen intermediates. "We have demonstrated that nitric oxide (NO), used widely as a signaling molecule in biological systems, induces dispersal and dissolution of *P. aeruginosa* biofilms at low, sublethal concentrations," Webb said. "Analogous to apoptosis in eukaryotes, this induction of programmed cell death by reactive oxygen or nitrogen intermediates followed by dispersal and biofilm sloughing appears widespread among bacteria."

These findings help to demonstrate how polymicrobial communities are modulated in the natural world, rather than simply being tolerated on the surfaces they occupy. However, they also indicate potential methods of dealing with unwanted biofilms on artifacts ranging from surgical implants to man-made structures in the sea. Several different applications of the furanones synthesized by *D. pulchra*, for example, are now being developed commercially. More broadly, Webb believes that, since nature modulates but does not totally eliminate biofilms, we should seek to control them by nonbiocidal strategies.

Cary Lambert of Nottingham University described work on another resident of polymicrobial biofilms, the gram-negative bacterium *Bdellovibrio*. A predator of other gram-negative bacteria, including pathogens of plants, humans, and other animals, it occurs in the soil and in marine and other aquatic environments. Vigorous flexing of its sheathed flagellum has allowed it to win the distinction of holding the world speed record for swimming by bacteria. *Bdellovibrio* uses type IV pili to locate and enter its prey by penetrating their outer layers, growing within the periplasm, and lysing the cell.

Lambert's talk was titled *Living Together while Being Eaten* because *Bdellovibrio* never destroys a population of prey entirely, even when let loose under laboratory conditions against inferior numbers of bacteria. Despite this limitation, which mathematical modeling is helping to explain, there are hopes of harnessing the organism as an alternative to antibiotics to combat gram-negative pathogens found in ulcers and burn wounds.

The effect of the use and abuse of antimicrobial substances over the last century on the nature of polymicrobial communities was discussed by Peter Gilbert of the University of Manchester. He used the results of microcosm studies to contrast the vigorous emergence of antibiotic-resistant bacteria with the absence of anything like the same trend for agents such as biguanides, triclosan, and quaternary ammonium compounds.

Products containing chemicals of this sort, he pointed out, are invariably deployed in situations with stable polymicrobial populations, such as soil or the skin, or on aesthetically clean, hard surfaces whose resident/transient species have broad susceptibility profiles. "Exposure to antibacterials will therefore kill/inhibit the growth of some strains and reduce the growth efficiency of others, yet leave many others unaffected," Gilbert said.

"While partially inhibited strains will be subject to a selection pressure toward less susceptible phenotypes, this will incur a fitness cost and a temporary loss of competitive efficiency.

"Climax communities generally resist the influx of new species, since adventitious arrivees must be able either to outcompete residents in terms of nutrient utilization and/or occupy vacant functional niches. During sublethal exposure to antibacterials, colonization resistance is lost and adapted strains must recompete for their position in the community. Invariably this battle is lost, with the effects of antibacterial use being a clonal expansion of preexisting, less susceptible strains with displacement of the susceptible ones."

Several papers highlighted new insights, coming from ribosomal sequence analysis, into the bacterial diversity of the human colon. In turn, this is helping researchers to predict the effects of dietary additives, such as probiotics, prebiotics, and nondigestible carbohydrates. Emma Woodmansey of Smith and Nephew Research Centre, York, United Kingdom, reported promising results from feeding trials designed to counteract adverse changes in the gut population. For example, as we age, our colons harbor rising numbers of facultative anaerobes and falling numbers of anaerobic lactobacilli and bifidobacteria. These shifts may result in increased putrefaction and greater susceptibility to disease.

There were also warnings in Edinburgh that bioremediation efforts often prove to be disappointing because specialized scavengers, developed in the laboratory, have to function not in isolation, but in the ecological networks of which they become a part. A team based at the University of Perugia in Italy described one project in which they inoculated *Botryosphaeria rhodina* in a site heavily contaminated with aromatic hydrocarbons and achieved superior results when the fungus worked in cooperation with the autochthonous microflora.

Whether applied to pristine, contaminated, or bioremediated soil or to the human gut in health or disease, the central message of the SAM conference may be summarized in the same words. While much can be learned about the behavior of specific isolates under laboratory conditions, understanding and modifying polymicrobial communities in the real world pose rather more formidable challenges.

References

1. **Sauer, K., A. H. Rickard, and D. G. Davies.** 2007. Biofilms and biocomplexity. *Microbe* **2:**347–353.

2. **Schaudinn, C., P. Stoodley, A. Kainoviæ, T. O'Keeffe, B. Costerton, D. Robinson, M. Baum, G. Ehrlich, and P. Webster.** 2007. Bacterial biofilms, other structures seen as mainstream concepts. *Microbe* **2:**231–237.

17 Our Most Abundant Coterrestrials

Could even the billions killed by diseases such as smallpox, influenza, and yellow fever be dwarfed as we recognize the real extent to which viruses fashion life on Earth and influence human affairs? A meeting held in 2007 in Edinburgh, Scotland, by the Society for General Microbiology indicated that a profound reappraisal of the ecology of viruses, now under way, could lead us to a conclusion at least as dramatic as this.

Just a few decades ago, the only viruses receiving serious attention were those we recognized as animal, human, or plant pathogens, plus the bacteriophages used for various purposes in the laboratory. Even worse, as a category, viruses were barely credited with the significance of true life forms. Today, we realize that they are the most abundant biological entities on the planet, with activities considerably more profound than the pioneer virologists ever imagined (see chapter 5). They not only define the composition and behavior of communities of bacteria, but play extremely important roles in biogeochemical cycling and in horizontal gene transfer. As our knowledge has deepened, viruses have become of concern not only to medical microbiologists and plant pathologists, but to all microbiologists, indeed to all biologists.

"Viruses are the major component of the biosphere, and cellular genomes are continuously visited by viruses/plasmids coming from a hidden viral reservoir of huge magnitude," said one of the Edinburgh speakers, Patrick Forterre of the Pasteur Institute in Paris, France. "Known viruses—the tip of the iceberg—exhibit a much greater diversity than cells in the nature of their genomic material and in their mechanisms of genome replication."

Mya Breitbart of the University of Florida, St. Petersburg, argued that we are only just beginning to understand the identities and diversity of viruses in the environment. "Addressing this issue is difficult because there are no conserved genes that are shared in all viruses—which must be cultured on hosts, many of which cannot be cultivated using standard techniques."

However, investigators have now begun to use metagenomic sequencing to study the identities and diversity of viruses in a variety of locations, including seawater, marine sediment, soil, and feces. "The majority of metagenomic sequences are not similar to those in the current databases, suggesting that environmental viruses represent the largest reservoir of unknown sequence space," Breitbart said. Moreover, mathematical modeling based on the distribution of overlapping sequence fragments from shotgun libraries suggests that viral communities are incredibly diverse, with over 10,000 viral genotypes in 1 kg of marine sediment and hundreds of thousands of viral genotypes in the world's oceans.

Against this backdrop, several specific projects reported in Edinburgh illustrate the range of systems within which the influence of viruses is being characterized for the first time. For example, Shah Faruque of the International Centre for Diarrhoeal Disease Research, Bangladesh, in Dhaka, Bangladesh, described recent work revealing how bacteriophage activity helps to account for the seasonal regularity of cholera epidemics. Following indications in 2001 to 2003 of the importance of lytic phages in the environment, Faruque explored the dynamics of their interaction with *Vibrio cholerae* during the entire course of an epidemic in Dhaka.

The results showed not only that the peak of the epidemic was preceded by high *V. cholerae* 10 prevalence in the environment, but that this was followed by high environmental levels of a phage, JSF4, to which the bacterium was sensitive. "The build-up to the phage peak coincided with increasing excretion of the phage in patients' stools. Thus in vivo phage

Animalcules: the Activities, Impacts, and Investigators of Microbes

amplification in cholera victims likely contributed to heightened environmental phage abundance.... Hence the collapse of the epidemic."

Meanwhile, at the University of Cambridge and the Wellcome Trust Sanger Institute, Hinxton, Cambridge, United Kingdom, Nicola Petty and colleagues have been extending our understanding of how prophages influence the virulence, evolution, and diversity of bacterial pathogens. While annotating the newly sequenced genome of the mouse pathogen *Citrobacter rodentium*, they identified six prophage-like elements, some of which proved to be active phages.

"Comparative genomic analysis showed that they had similarities to Mu and P2 phage families, while some appeared to be conserved among other enteric pathogens," Petty said. "Other prophage regions seemed specific to *R. rodentium*, and many of the prophages carry 'cargo' genes possibly involved in pathogenicity. Several of the prophage-like elements are predicted to have disrupted core functions due to their insertion in the host genome. For example, insertion into flagella operons may be associated with *C. rodentium*'s lack of motility."

Mathias Middelboe of the University of Copenhagen, Copenhagen, Denmark, has evaluated viral activity in a very different environment, that of marine sediments, which comprise one of the planet's largest ecosystems. While recent pelagic research has pointed to the influence of viruses on pelagic microbial mortality, diversity, and biogeochemical cycling, these phenomena have so far been little studied.

Recent work, however, has confirmed that viral communities are highly diverse and dynamic players in benthic ecosystems. "One cm^3 of surface sediment contains 10^8 to 10^{10} viruses, a 10- to 100-fold-higher density than in the upper water column," said Middelboe. "So with estimated turnover rates of 0.5 to 5 days, viruses may constitute a significant loss factor for benthic bacteria—up to 40% of bacterial production. Indeed, accumulating data indicate that viruses are important, integrated components of the benthic biosphere."

Cyanobacteria have been recognized for many years as key members of phytoplankton populations, in both the seas and freshwater. Now, increasing evidence is showing the importance of cyanophages, and of lateral gene transfer, in regulating the structure and evolution of cyanobacterial communities. Because little is known of these phenomena in freshwater, Li Deng and Paul Hayes of the University of Bristol in the United Kingdom established a study in the Cotswold Water Park in the United Kingdom and Lake Zurich in Switzerland.

From those sites, they isolated 35 different cyanophages capable of infecting bloom-forming cyanobacteria of the genera *Microcystis, Anabaena,* and *Planktothrix.* "The isolates encompassed a variety of morphological types, including the first filamentous cyanophage. Some could infect all three genera of cyanobacteria," said Deng. "The ability to infect a wide range of host taxa extends the potential reproductive period for lytic propagation and has implications for the transfer of genetic information between deeply separated cyanobacterial lineages."

Another Edinburgh speaker, Johanna Laybourn-Parry of the University of Keele in the United Kingdom, described work she has been doing with colleagues at the University of Lund in Sweden on lakes, ranging from hypersaline to freshwater, in Antarctica. Their studies indicate that viruses may play a significant role in the cycling of carbon in these microbe-dominated systems.

A challenging issue in the hinterland of all of these papers was the much broader question of the nature and origin of viruses. As Patrick Forterre pointed out, the discovery of unusual viruses in *Archaea* and giant varieties in recent *Eukarya*, combined with progress in comparative genome and structural biology, have brought this question to prominence again.

"Present cellular genomes and mechanisms of DNA replication could be a subset of those invented in the primordial virosphere of DNA viruses infecting RNA cells," he said. "If true, viruses may have played a major role in the transition from ancient cellular RNA genomes to the DNA genomes of modern cells, and possibly in the establishment of the first cellular domains of life."

Think of it. Beyond our historical perspective on viruses solely as agents of death and disease, beyond even our emerging awareness of their manifold global roles, viruses could have "invented" DNA.

Animalcules: the Activities, Impacts, and Investigators of Microbes

18 *Helicobacter* from the Seas?

During much of the 20th century, medicine advanced on two parallel tracks: the progressive refinement of therapeutics and the increasing sophistication of surgery. From time to time, novel drugs replaced the surgeon's scalpel, but rarely did the reverse occur.

Now things are less simple. Exquisitely skillful operations supplant imperfect therapies, while in the reverse direction, traditional surgical procedures give way to ingeniously targeted pharmacology.

Peptic ulcers and coronary heart disease are two outstanding examples. Since Barry Marshall and Robin Warren's work on *Helicobacter pylori* in Perth, Australia, in 1982, gastrectomies for intractable peptic ulceration have been largely consigned to history. On the other hand, the type of patients who even at that time were struggling to combat their angina with the aid of glyceryl trinitrate tablets now receive a coronary bypass. My father, who suffered a fatal heart attack at the age of 56 after many years of angina, would have been a prime candidate for the operation.

What unites these contrary developments is their sheer unlikelihood in the opinion of the vast majority of experts at the time. I recall eminent physicians in the 1960s arguing that, despite limited experimental success in revascularizing the ailing myocardium of dogs, it was totally unrealistic to expect any such benefit for human patients. If coronary bypass surgery were ever to succeed, it would require huge surgical teams, be immensely costly, and help only a tiny minority of patients. Today, it is a routine procedure all over the world, transforming the quality of life for millions and saving countless lives.

Even more bizarre, for both microbiologists and surgeons, was the notion that peptic ulcers might be attributable to an infectious agent. Indeed, skepticism and outright disbelief persisted for some time after Marshall and Warren's initial report. Today, *H. pylori* is recognized as colonizing the gastric mucosa of about half the world's population. Infections are mostly asymptomatic, but in 10 to 20% of carriers, chronic infections are associated with conditions ranging from ulceration to adenocarcinoma.

Hence, pharmacopeias of the early 21st century contain potions— antibiotics for peptic ulcer disease—that would have been considered absurd even by pharmacopeia compilers of the late 20th century. The most favored formula includes clarithromycin and amoxicillin plus a protein pump inhibitor, lansoprazole, to damp down acid production. I had a course myself some months ago.

In 2 decades, we have learned an astonishing amount about *H. pylori*, its pathophysiological role, its effects on cell signaling and cell cycle regulation, its genome sequence (delineated in 1997), its population structure, and much more besides. However, a rather substantial problem remains. We do not really know where *H. pylori* comes from.

Our ignorance is not total. Most infections seem to be acquired during childhood, probably through the fecal-oral route, but published claims regarding the source of the organism have ranged from water supplies to milk and seafood. As we approach a quarter century since that historic breakthrough in Perth, *H. pylori*'s precise route and mode of transmission remain unproven.

All the more reason, then, to welcome some important clues that have emerged from studies on the Italian side of the Adriatic Sea, which lies between Italy and the Balkan Peninsula. Work there by Luigina Cellini and colleagues suggests that a significant reservoir for the organism is seawater, where it occurs both in a free-living form and in association with plankton. The researchers believe they may have located an important source of human infections.

Cellini, who works at the G. D'Annunzio University in Chieti, Italy, conducted the survey in collaboration with investigators at the University of Rome; the Istituto Superiore di Sanità, also in Rome; and the Agency for the Protection of the Environment in Pescara, Italy. They reported their findings in the *Journal of Applied Microbiology* (1).

The paper described a simple, quick, multistage DNA preparation method that the Italian group used to search for *H. pylori* in seawater. The nested PCR specifically amplifies a highly conserved region of the phosphoglucosamine mutase *glmM (ureC)* gene, which is unique and essential for the organism to grow. Previous work has shown that it boosts the sensitivity of detection of *H. pylori* in samples containing both prokaryotic and eukaryotic cells, in addition to organic impurities.

Every month for a year (from May through April), Luigina Cellini and coworkers recovered water from a sampling station at a depth of 5 m about 500 m from the Italian Adriatic coast. They screened them, not only for their target organism, but also for others, such as enterococci and fecal coliforms, whose presence is normally assessed in determinations of the microbiological quality of recreational waters. Appropriate filters were used to enumerate free bacteria and plankton-associated *H. pylori* DNA.

Seven of the 12 monthly samples contained *H. pylori* either free or bound to planktonic organisms. During the summer (when coastal waters usually have a higher level of bacterial pollution), the water carried the free organism. In November, December, and March, however, it tended to be associated with plankton.

"Our data also show a significant presence of *H. pylori* linked to Copepods and Cladocerans," the researchers wrote. "In fact *H. pylori* cells bound to plankton were detected either in summer months or in November/December during the blooming of Copepods and in March when Cladocerans are present in greater numbers. Therefore, we suppose that zooplankton organisms represent a sort of protected niche for survival of *H. pylori*." The discovery of the bacterium attached to plankton is particularly notable in light of the role of plankton in the seafood chain, and thus the possible significance of seafood in the spread of *H. pylori* to humans. More generally, these findings from the Adriatic indicate that polluted coastal marine environments may provide a significant reservoir of the organism to infect swimmers and others using those waters for work or pleasure.

Meanwhile, 250 miles to the south of Cellini's sampling station, marine research has provided further disquieting news about the health aspects of pathogens living in association with plankton. Working in the Straits of Messina, between the southernmost tip of mainland Italy and Sicily, microbiologists from the University of Messina found evidence (2) that the colonization of zooplankton by organisms capable of causing human disease is a widespread phenomenon.

This survey had several purposes, including an assessment of the occurrence of species of *Campylobacter*, *Vibrio*, and other genera in Italy's coastal waters, together with comparisons of free-living bacteria and those associated with zooplankton and of plankton-bound organisms with selected pathogens. A variety of enrichment and selective techniques were used.

One of the most significant findings was that not only *Vibrio* and *Aeromonas* spp. were linked with zooplankton (as reported previously), but so too were *Escherichia coli*, enterococci, and *Campylobacter* and *Arcobacter* spp. (agents of human diarrhea). An abundance of both free-living and plankton-associated *E. coli* strains and enterococci confirmed that the Straits of Messina were indeed seriously polluted.

Together with the findings in the Adriatic Sea, these results indicate that potentially pathogenic organisms living in close association with zooplankton have considerable epidemiological (and ecological) implications. There is, of course, no reason to believe that the situation is unique to the coastline of Italy. However, this is some of the most convincing evidence of its sort yet adduced. All credit is due to the government's research ministry in Rome, which financed the work.

References

1. **Cellini, L., A. Del Vecchio, M. Di Candia, E. Di Campli, M. Favaro, and G. Donelli.** 2004. Detection of free and plankton-associated *Helicobacter pylori* in seawater. *J. Appl. Microbiol.* **97:**285–292.

2. **Maugeri, T. L., M. Carbone, M. T. Fera, G. P. Irrera, and C. Gugliandolo.** 2004. Distribution of potentially pathogenic bacteria as free living and plankton associated in a marine coastal zone. *J. Appl. Microbiol.* **97:**354–361.

Animalcules: the Activities, Impacts, and Investigators of Microbes

19 Selective Agencies

Looking back half a century to the emergence of bacterial resistance as a clinical nuisance, one is reminded just how straightforward things seemed then. There were difficulties, but they were susceptible to an elementary solution. Simply switching to another could circumvent insensitivity to one antibiotic. As knowledge replaced naiveté, however, we realized that the entire resistance spectrum of a multiply resistant organism could be selected for by any one of that portfolio of drugs.

Today, despite some successes in combating the problem by more judicious prescribing and other restrictions, we are becoming aware of a further dimension. We now recognize that agents other than antimicrobials can foster selective pressures. Two such agents that have fallen under suspicion are as disparate as could be imagined: a commonly prescribed tranquilizer (diazepam) and a metal (copper) used by farmers throughout the world as a fungicide and bactericide.

The year 2004 brought the first evidence that everyday domestic products can induce the *mar* operon, which is essential for many bacterial species to adopt a multiple-antibiotic-resistant phenotype. Peter Gilbert and coworkers at Manchester University in the United Kingdom found that several seemingly innocuous culinary herbs, for example, induced *mar* expression in *Escherichia coli* (2). While the practical implications of this finding remain to be clarified, the discovery underlines the fact that much remains to be learned despite decades of intensive scrutiny of antibiotic resistance.

The last few years have also seen reports that chemotherapeutic drugs, such as clofibric and ethacrynic acids, as well as chemicals, including sodium salicylate and sodium benzoate, induce the Mar phenotype in *E. coli* by activating the *marRAB* operon. This led Maria Tavío and colleagues at the Universities of Las Palmas and Barcelona in Spain and the University of L'Aquila in Italy to investigate diazepam as a resistance inducer. They were encouraged to do so by evidence that the tranquilizer reduced the survival of both humans and mice with severe bacterial infections. This seemed to occur in part because the drug impaired the immune response.

Tavío's group decided to compare diazepam with sodium salicylate and sodium benzoate as known *marRAB* inducers. The resulting data showed very similar effects on the outer membrane protein expression, active efflux, and MICs of various antimicrobial agents in *Klebsiella pneumoniae* strain KP1A02 and *E. coli* strain Ag100 (3). "These results suggest that diazepam concentrations equal to or twice adult dosage might induce the expression of MarA in *E. coli* strains or RamA in *K. pneumoniae* strains," they concluded.

In addition to its capacity to interfere with the operation of the immune system, therefore, the commonly used tranquilizer may also induce the Mar phenotype, leading to a doubly intractable infection. This prospect prompted Tavío and her collaborators to point out the possible risk of the development of resistance in patients concomitantly treated with diazepam for insomnia or other problems and with antibiotics to combat an infection.

This is a worrisome scenario. Even more disquieting, however, is another addition to our growing catalogue of agents that may create selective pressures favoring the proliferation of organisms insensitive to antimicrobials. It has received attention through the work of Ole Nybroe at the Royal Veterinary and Agricultural University in Frederiksberg, Denmark, and that of the Danish Institute for Food and Veterinary Research in Copenhagen, Denmark. Their studies suggest that copper, a metal that is introduced both consciously and accidentally into farmland throughout the world, can have a significant effect on the selection of antibiotic resistance among bacteria in the soil.

Animalcules: the Activities, Impacts, and Investigators of Microbes

Copper is used in several forms as an agricultural fungicide and bactericide but also occurs in the sewage sludge and animal manure that farmers in many places spread on their fields as fertilizer. The metal persists in topsoil with an estimated half-life of hundreds or thousands of years. Inevitably, this persistence is accompanied by the emergence of copper-tolerant strains of soil bacteria.

However, it was not these considerations alone that alerted the Danish team to a possible problem. The real trigger was the accumulation in recent years of signs that terrain contaminated with copper contains, in addition to copper-tolerant bacteria, a higher-than-average proportion of organisms insensitive to antimicrobials. These observations raised the possibility that metal pollution also exerts some form of indirect selection favoring the proliferation of drug-resistant bacteria (1). Unlike many other pollutants, copper is not degraded, so the phenomenon would be long lasting.

Ole Nybroe's group decided to establish a field experiment to quantify the impact of copper supplementation on the frequencies of both copper resistance and antibiotic resistance patterns in indigenous soil organisms. They used a randomized block design for the trial on a sandy loam field, which included three plots treated with copper sulfate (at a concentration close to the 140-ppm European Union limit for soil fertilized with sewage sludge) and three untreated plots. The entire field received only inorganic (nitrogen, phosphorus, and potassium) fertilizer throughout the study and for 5 years beforehand. Soil samples were taken from both types of plot 21 months after the commencement of the experiment. They were sieved and stored in the dark for up to 10 months after being pooled with other samples from the same plot.

The investigators screened their isolates for resistance to ampicillin, chloramphenicol, nalidixic acid, olaquindox, streptomycin, sulfanilamide, and tetracycline. Since there is no validated procedure to determine the drug resistance of uncharacterized soil bacteria, they chose concentrations of antibiotics for the ability to inhibit the growth of ca. 80% of 76 isolates included in a preliminary experiment.

Predictably, perhaps, copper supplementation significantly increased the frequency of Cu-resistant isolates, more than 95% of which proved to be gram negative. However, they also had significantly higher resistance to ampicillin, sulfanilamide, and multiple (three or more) antibiotics than Cu-sensitive gram-negative isolates. Moreover, Cu-resistant gram-negative organisms from Cu-treated plots had a significantly higher incidence of resistance to chloramphenicol and multiple (two or more) antimicrobials than corresponding isolates from control plots.

It does seem, therefore, that copper added to the soil not only selected for Cu-resistant organisms, but also coselected for antibiotic resistance. The Danish Agricultural and Veterinary Research Council was wise to sponsor the study, which may, however, have far wider implications.

There is a rather quirky sidebar to this story. One of the oddities of the British craze for organic food in recent years is that organic farmers, while rejecting modern, safe, thoroughly tested pesticides, are permitted to spray their crops with copper sulfate. Indeed, they lobbied the European Union so effectively that they have been allowed to continue using it despite a ban that was scheduled to come into effect in 2004. Though long employed as a constituent of Bordeaux mixture to combat *Phytophthora infestans* (the agent of the Irish potato famine at the end of the 19th century), copper sulfate was due to be prohibited on safety grounds. It has, for example, caused well-documented cases of liver damage among vineyard workers. Nevertheless, this potent hepatotoxin, which is persistent in soil, continues to be sprayed widely in the ostensible interests of naturalness.

Could it be that well-intentioned but misguided organic farmers are not only deploying a rather nasty metallic poison in the ostensible pursuit of pristine purity, but also adding to environmental pressures favoring the proliferation of antibiotic-resistant bacteria?

References

1. **Alonso, A., P. Sánchez, and J. L. Martínez.** 2001. Environmental selection of antibiotic resistance genes. *Environ. Microbiol.* **3:**1–9.

2. **Rickard, A. H., S. Lindsay, G. B. Lockwood, and P. Gilbert.** 2004. Induction of the *mar* operon by miscellaneous groceries. *J. Appl. Microbiol.* **97:**1063–1068.

3. **Tavío, M. M., J. Vila, M. Perilli, L. T. Casañas, L. Maciá, G. Amicosante, and M. T. Jiménez de Anta.** 2004. Enhanced active efflux, repression of porin synthesis and development of Mar phenotype by diazepam in two enterobacteria strains. *J. Med. Microbiol.* **53:**1119–1122.

20 Natural Disaster Microbiology

Epidemics of cholera, dysentery, typhoid fever, and other waterborne plagues have long been feared as possible sequelae to natural disasters, such as the cyclone in Burma in May 2008. More recently, heightened dangers from measles and hepatitis viruses have become apparent in these situations. Now, another significant problem is attracting attention—that of polymicrobial infections with multiply resistant organisms heavily inoculated into the body during trauma. Indeed, it begins to seem that the "classical" threats posed by enteric pathogens have been overemphasized while those coming from other organisms have been underemphasized.

It is still commonplace to read that outbreaks of waterborne communicable disease in the wake of a natural disaster may claim more lives than the event itself. The risk unquestionably exists, yet the hard evidence provides a rather different picture. After conducting a 10-year review of the literature, Didier Pittet and his colleagues pointed out (2) that such infections are not usually the major cause of mortality in disaster victims. Deaths occur mainly through trauma. There were no severe communicable-disease outbreaks in the aftermath of the 2004 tsunami or hurricane Katrina in 2005. What the records show are cases of tetanus, measles, and nonspecific diarrhea, together with a norovirus outbreak among evacuees in a temporary shelter facility in Texas.

As Gretchen Vogel reported in *Science* (4), it was a 1990 study by Ronald Waldman of Columbia University and Michael Toole, now of the Burnet Institute in Melbourne, Australia, that first highlighted the importance of measles following disruptions to normal living. They pointed out that the disease pushed the mortality rate of children in Ethiopian refugee camps up to 60 times its normal level. Despite the fact that measles was known to be a potential killer, especially among malnourished children, its potentially catastrophic role in this type of situation came as rather a surprise to many aid agency workers.

Waldman and Toole's work triggered a reassessment by the World Health Organization of the most prudent priorities to be adopted when health services assist the population of a stricken area in coping with the threat of communicable diseases. Thus, when a relief worker rang WHO officials in Aceh, Indonesia, on 8 January 2005, 2 weeks after the tsunami, to report measles in an affected village, help was immediately available. By that afternoon, 1,000 people in the area had been vaccinated.

Didier Pittet, who is based at the University of Geneva Hospitals in Switzerland, has been working with colleagues there and at two hospitals in France to build up a picture of the prevalence of multiply resistant microorganisms among survivors of natural disasters. Their survey, built upon an extensive Medline and PubMed search for papers on infections in disaster victims, focused principally on the aftermath of the tsunami.

One of the most striking findings was that pathogens in repatriated patients were often resistant to several antibiotics. "Moreover, contrary to classical orthopaedic infections, where *Staphylococcus aureus* is the main pathogen, Gram-negative rods and extended-spectrum beta-lactamase (ESBL)-producing bacteria predominated over Gram-positive pathogens in tsunami victims, according to reports from Thailand, Germany, Finland, Sweden, Italy, Australia and Switzerland," Pittet and his coworkers reported. "These infections were polymicrobial, including atypical fungi and mycobacteria.

"Among nine patients transferred to our institution for treatment or multiple fractures, seven were colonised (*methicillin*-resistant *S. aureus*, ESBL-producing *Escherichia coli*) or infected (multi-resistant *Acinetobacter baumannii*, multi-resistant *Stenotrophomonas maltophilia*, *Sce-*

dosporium apiospermium, Alcaligenes xyloxidans, Enterococcus faecium, Pseudomonas aeruginosa, Nocardia africanum, Mycobacterium chelonae) by multi-resistant organisms causing severe infections such as cerebral abscesses or spondylodiscitis. Almost every unusual microorganism found in clinical specimens proved to be a potentially infective agent."

The notion that these organisms originated in the environment is supported by many reports from other centers. Several have described species of *Aeromonas* and *Vibrio*, clearly related to the marine environment, in victims of the tsunami and Hurricane Katrina. Also, the incidence of melioidosis tends to rise after natural disasters, while victims of events ranging from the Marmara earthquake in Turkey in 1999 to tornadoes in Georgia and Alabama have developed infections with gram-negative rods.

"Atypical fungal and mycobacterial infections were often reported in polymicrobial infections in immunocompetent patients, whereas a high incidence is normally encountered only in transplant recipients," the Swiss-French group reported. "Since these fungi are ubiquitous in soil and water, traumatic inoculation at the disaster location is presumably the origin, although some reports suggest that extensive water damage of hospitals could increase the likelihood of mold contamination."

As to why many of the posttsunami isolates were insensitive to several antibiotics, Pittet and his collaborators point out that soil is known to be a reservoir for the development of bacterial resistance to clinically relevant antimicrobials. "Antibiotic-resistant bacteria have been isolated from virtually every environment and region, including South-East Asia, where they are not part of the normal skin flora." Another argument for an environmental origin of both the organisms and their multiple resistances was the low prevalence of one organism that certainly is associated with the health care environment—methicillin-resistant *Staphylococcus aureus*—compared with gram-negative rods.

The importance of soil as a reservoir of resistance to antimicrobial agents may have been underlined by recent work by Gautam Dantas and coworkers at Harvard Medical School, Boston, MA, and Harvard University, Cambridge, MA. In the first systematic study of its sort, they isolated hundreds of soil bacteria that were able to grow on antibiotics as the sole source of carbon (1). Of 18 antibiotics tested, representing eight major classes of natural and synthetic origin, 13 to 17 supported the growth of clonal bacteria from each of 11 diverse soils.

Dantas and his colleagues were surprised to find that the bacteria subsisting on antibiotics were phylogenetically highly diverse, with many related to human pathogens. "Furthermore, each antibiotic-consuming isolate was resistant to multiple antibiotics at clinically relevant concentrations," they wrote. "This phenomenon suggests that this unappreciated reservoir of antibiotic-resistance determinants can contribute to the increasing levels of multiple antibiotic resistance in pathogenic bacteria."

The Swiss-French findings are further components in a still-evolving view of the impact of natural disasters on the transmission of communicable diseases. On one hand, the emphasis is shifting away from a focus on the fear of waterborne epidemics traditionally thought to be associated with disasters. We now realize that these do not necessarily occur, even, paradoxically, after large-scale floods. "In the past three decades, epidemics of water-borne illnesses, such as cholera and Shigella dysentery, have been uncommon after floods and natural disasters," wrote Michael VanRooyen and Jennifer Leaning in the *New England Journal of Medicine* (3). "They are quite common, however, in large displacement centers and refugee camps. Other common communicable diseases such as acute respiratory infections and measles result in high mortality in populations under stress—particularly among children younger than five years of age—when they are living in large refugee camps.... It is not the disaster but the artificial, crowded communities ... that serve as a substrate for the spread of communicable diseases."

It is one thing to recognize a new order of priorities in dealing with infectious-disease threats in the aftermath of a natural disaster. As VanRooyen and Leaning indicate, the logistical challenge of meeting those challenges is more complex. I wonder, for example, about the practicality of some of Pittet's suggestions that patients should not share lavatories or meal facilities with other patients and should only be transported in aircraft fitted with high-efficiency particulate air filtration.

Such practicalities aside, the microbiology of disaster relief is clearly much more soundly grounded than it was 20 years ago.

References

1. **Dantas, G., M. O. A. Sommer, R. D. Oluwasegun, and G. M. Church.** 2008. Bacteria subsisting on antibiotics. *Science* **320:**100–103.

2. **Uçkay, I., H. Sax, S. Harbarth, L. Bernard, and D. Pittet.** 2008. Multi-resistant infections in repatriated patients after natural disasters: lessons learned from the 2004 tsunami for hospital infection control. *J. Hosp. Infect.* **68:**1–8.

3. **VanRooyen, M., and J. Leaning.** 2005. After the tsunami—facing the public health challenges. *N. Engl. J. Med.* **352:**435–438.

4. **Vogel, G.** 2005. Indian Ocean tsunami. Using scientific assessments to stave off epidemics. *Science* **307:**345.

21 Foot-and-Mouth Folly?

If any government were to arrange a perfect scenario for a highly infectious (and possibly highly virulent) virus to wreak havoc on that country's population of cows or other farm animals, it would need to take two steps. First, it would have to resolve not to protect its national herd(s) by immunization. Indeed, it would ban farmers from doing so. This would leave the animals totally vulnerable to the infection.

Secondly, the government would have to arrange its agricultural practices so that susceptible animals were moved frequently and widely between farms, holding centers, markets, and abattoirs. This would ensure that, if the virus were to enter the country, it would be disseminated widely and efficiently to other unprotected livestock.

These two conditions would create the ideal setting for "a disaster waiting to happen." However effective the country's other precautions to exclude the virus, its eventual accidental (or deliberate) introduction from outside would be inevitable. The danger would be all the greater in the case of an organism that circulated freely in other parts of the world.

This is precisely the scenario that was allowed to develop ahead of the foot-and-mouth disease (FMD) epidemic that began in Britain in 2001. Its predictable consequences then emerged. The first indication of trouble was on 20 February 2001, when scientists at the World Reference Laboratory for FMD at Pirbright, Surrey, United Kingdom, found that samples sent from an abattoir in Essex were positive for FMD virus. The organism proved to be a pandemic serotype A virus (now named PanAsia virus), which has spread in recent years from India into Saudi Arabia and thence into Europe, China, Russia, Mongolia, South Africa, and several other regions.

United Kingdom investigators traced the virus detected on 20 February to infected pigs on a farm in the north of England. Thereafter, it moved rapidly, and within 3 weeks, 200 outbreaks had been confirmed around the country. By then, the Ministry of Agriculture had taken extensive actions based on its policy of containment and slaughter. Burning pyres of dead livestock appeared throughout England; movements of animals and people were severely curtailed; meat exports were terminated; zoos, parks, and farms closed to the public; major sporting events were canceled; and meat supplies ran short in supermarkets. By 14 March, 120,000 animals had been slaughtered.

This entire calamity could have been foreseen and perhaps circumvented if efforts had been made to assess FMD from an ecological perspective. Instead, Britain had followed a policy uncomfortably reminiscent of many previous mistakes in dealing with infectious diseases around the world. Examples range from the failure of dam builders to realize they were creating fresh breeding grounds for mosquitoes to that of refrigeration engineers in opening up new niches for low-temperature opportunists, such as *Listeria monocytogenes*.

FMD causes an acute vesicular condition of pigs and wild and domesticated cattle, sheep, goats, deer, and other ruminants. Although the mortality of young animals can be high, the disease's main effect in older animals is to impair meat and milk production. For these reasons, many countries do not use either vaccination or containment and slaughter. They simply live with the infection and tolerate its significant economic cost.

In this sense, the threat of FMD hardly justifies the apocalyptic prose used by United Kingdom journalists when the epidemic erupted in February 2001. Nevertheless, the virus is highly transmissible, arguably the most contagious organism on Earth. Infected pigs excrete the virus especially prolifically through a rich aerosol of virus particles, which can be carried by the wind. It also spreads passively but swiftly through the movement of animals (including humans and horses) and through products such as meat and milk.

The European policy on FMD changed in the early 1990s. First, countries such as France, Germany, and Italy, which had previously immunized livestock, phased out the practice. They resolved to control any future outbreaks by killing and destroying all affected animals (plus, in some cases, in-contact animals) combined with emergency "ring vaccination" in surrounding areas. Britain, Greece, and the Republic of Ireland, which had not hitherto vaccinated animals, continued to use slaughter alone.

Shortly after these changes, the European Community agreed on a unified approach. Following a cost-benefit comparison, immunization was terminated throughout the Community in favor of reliance on containment and culling in the event of an epidemic. Herd immunity has declined accordingly.

As an island, Britain has always felt more secure from the threat of FMD than countries with land borders. Nevertheless, the virus did enter the United Kingdom in the past, with severe consequences. In 1967, Argentinean lamb was the suspected source when an outbreak began in Oswestry and spread with frightening speed; 400,000 farm animals eventually had to be killed, at a cost to the Ministry of Agriculture of £34 million. The total bill came to $256 million in lost sales, slaughtering costs, and compensation paid to farmers (*The New York Times*, 22 February 2001).

There was a lucky escape in 1981, when FMD broke out in Jersey and the Isle of Wight off Britain's south coast but did not reach the mainland. However, that incident vividly illustrated the airborne transmissibility of the virus, which almost certainly arrived in a plume from infected pigs across the English Channel in France.

The 1967 outbreak, though costly, remained relatively localized. The major difference in 2001 was the very rapid spread of the virus throughout the country. This reflected changes in agriculture, which meant that farm animals were being moved often and over large distances, greatly increasing the risk of wide-scale dissemination. For example, many local abattoirs had closed down in recent years, necessitating long journeys to centralized facilities.

Why did Britain, and indeed the whole of the European Union, reject immunization? One argument was that, once a country vaccinates its

Animalcules: the Activities, Impacts, and Investigators of Microbes

cows and other susceptible animals, it loses its status as an FMD-free zone, yet this was the situation which eight European countries found quite satisfactory at one time.

The other two arguments were that FMD vaccines, though effective, are costly and imperfect in several ways. In particular, they must be given at less than yearly intervals in order to provide satisfactory antibody titers. We do, however, control many other communicable diseases with imperfect vaccines, and vaccine technology has progressed considerably over the years.

Even the keenest advocates of the crude, dispiriting policy of containment and slaughter conceded that Britain, and indeed Europe, had created a worrisome scenario. Its twin elements were national herds totally vulnerable to the most infectious disease in the world and widespread livestock movements to facilitate dissemination of the virus if and when it was introduced from outside.

The two books I recall most vividly from my student days are Macfarlane Burnet's *Biological Aspects of Infectious Disease* and René Dubos's *Mirage of Health*. Both taught me the importance of seeing communicable diseases, not from a narrow medical or veterinary viewpoint, but from a broad ecological perspective. Four decades later, I wonder what Burnet and Dubos would have made of the 2001 outbreak of FMD in Britain. They would certainly have seen it coming.

When a Royal Society inquiry into the outbreak reported in 2002, it called for an international research effort to develop a vaccine conferring lifelong sterile immunity against all strains of the virus and made 34 different recommendations intended to prevent any repetition of the calamity of 2001. Little can the committee have guessed that some of its recommendations, in particular, those on the need for rapid crisis management in the event of another outbreak, would be required and, thankfully, shown to be effective as early as August 2007. That was the month when, to the embarrassment of all concerned, the government-financed Institute for Animal Health at Pirbright in Surrey was the center of another FMD outbreak. A bizarre combination of neglected, leaky pipes and England's unusually wet summer meant that the virus came either from the institute or from a private company on the same site. The cost of the measures established to contain the outbreak and its associated economic impact? Over $100 million.

22 Ecology Lessons

If we think of Great Britain as one gigantic microbiology laboratory, it is remarkable that three major "natural experiments" that occurred there in the early 21st century have each pointed to the same conclusion, that is, the importance of addressing communicable diseases from a wide, ecological standpoint rather than on the simplistic basis of encounters between specific pathogens and their hosts. The natural experiments were the 2001 foot-and-mouth disease epidemic; the decline in the uptake of measles, mumps, and rubella (MMR) vaccine, presaging the likely return of outbreaks of measles after its virtual eradication; and the arrival of H5N1 influenza virus in Scotland in 2006.

The theoretical foundations of ecology-based epidemiology were, of course, laid long ago, when W. H. Hamer established the "mass action principle" according to which the net rate of spread of an infection is proportional to the product of the densities of susceptible and infectious individuals in a population (1). W. O. Kermack and A. G. McKendrick (4) then showed that a few infectious people, entering a community of susceptible individuals, will trigger an outbreak only if the density or number of susceptibles is above a certain threshold.

In practical terms, however, the spectacular successes achieved by "magic bullets" (the apparent conquest of tuberculosis, for example) and vaccination (as in smallpox eradication) during the 20th century seemed to eclipse theory and ecology. Not only have public and medical perceptions of communicable disease remained harnessed to a pathogen/patient model, politicians, administrators, and funding bodies, too, have paid insufficient attention to the value of ecological and mathematical analysis, especially in laying contingency plans during times when particular infections are in abeyance.

Consider the foot-and-mouth disease epidemic that began on a farm in northern England in February 2001. Over the ensuing weeks, the virus spread far and wide, leading to the eventual slaughter of 4.2 million cows and sheep. The total cost attributable to interference with agriculture and the food supply was about £3.1 billion, and cleanup costs and compensation for slaughtered animals totaled another £2.5 billion.

Early priorities in the United Kingdom's response to the outbreak were the need to determine the average number of secondary cases triggered by one infected animal and to understand the spatial spread of the epidemic. However, immediate control measures were inadequate. When infectious-disease ecologists came in 2 weeks later, they quickly determined that each affected farm was likely to infect three to five more. More stringent measures were then adopted to identify infected and adjacent farms immediately when the virus was detected. This led to rigorous controls on animal movements and to the culling of herds.

It was the perspective of ecological theory that revealed a glaring deficiency in the United Kingdom's initial capacity to predict the spatial dynamics of the infection. This was the imperfect information, logged in a national database, concerning the sizes and geographical distribution of sheep and dairy farms. Only when further data were gathered and collated could investigators make confident predictions, having identified both local transmission of the virus by aerial plumes between neighboring farms and longer-distance dissemination through the movement of farmers (and veterinarians) and their vehicles.

"The UK foot-and-mouth outbreak provides some salutary lessons for public policy and future infectious disease outbreaks," wrote Katherine Smith of the University of California, Santa Barbara, and colleagues (7). "Retrospective calculations on the time course of the epidemic suggest that the initial 2-week delay may have eventually led to a doubling of cases and of the number of cattle and sheep herds culled. All of this suggests that the ecologists should have been brought in earlier."

The need to use mathematical models of epidemiology in "peacetime," i.e., in the absence of this infection or, indeed, others, was also highlighted by Mark Shirley and Stephen Rushton of the University of Newcastle upon Tyne in the United Kingdom (6). Their own theoretical study of the 2001 epidemic "emphasises the need to understand the contact pattern of susceptible populations before embarking on any strategy for disease control, which means that populations at risk from disease need to be characterised topographically before an outbreak occurs."

Ironically, my second United Kingdom scenario is the threat of renewed epidemics of the very condition, measles, whose study has provided much of our current understanding of infectious-disease dynamics. Many parameters, ranging from seasonality to the age structure of a population, affect the occurrence of measles outbreaks. However, the crucial factor that determines their outcome is again the reproductive number—the mean number of secondary infections per primary infection.

This key parameter is roughly proportional to the nonimmunized fraction of the population. In a community where the reproductive number is less than one, measles occurs solely as outbreaks when infected individuals from elsewhere introduce the virus, but if the number approaches one, then there are likely to be large outbreaks. Finally, the infection will become endemic if the reproductive number exceeds one.

Thanks to a needless and baseless scare over MMR vaccine, leading many parents to reject immunization for their children, the United Kingdom population has been approaching this critical point. While the World Health Organization was celebrating a 60% fall in measles cases in Africa since 1999, a decline in the percentage of immunized youngsters was pushing Britain in the opposite direction.

The country had already witnessed the first national outbreak since the establishment of mass vaccination (2). This occurred in a community devoted to the ideas of Austrian mystic Rudolf Steiner, whose adherents reject immunization because they believe that children benefit from the illness itself. Even more disquietingly, an analysis of recent outbreaks by Vincent Jansen and colleagues at Royal Holloway University of London and elsewhere (3) indicated that the United Kingdom situation had come close to criticality in recent years.

"If the current low level of MMR vaccine uptake persists in the UK population, the increasing number of unvaccinated individuals will lead to an increase in the reproductive number and possibly the re-establishment of endemic measles and accompanying mortality," they wrote. "In their attempts to avoid the perceived risk associated with vaccination, parents' behaviour collectively results in a substantial increase in the real risk of exposure to measles."

My third and final example is the national alarm that erupted after a dead swan carrying H5N1 influenza virus was found in the Scottish fishing village of Cellardyke on 7 April 2006, precipitating fears of an economically disastrous outbreak among chickens and, indeed, a human epidemic. Suddenly, virologists began to develop a deep interest in the migratory patterns of wild birds, farmers started to liaise with ornithologists, and wildlife conservationists realized they needed to understand the meaning of words such as neuraminidase and hemagglutinin.

The scenario underlying the anxieties has, of course, been familiar for years, but only in its broadest outlines. Ecological detail and mathematical analysis have been largely missing. "The role of wild birds in the spread of influenza H5N1 remains speculative and the ecology of influenza A viruses in nature is largely unstudied," wrote David Melville and Keith Shortridge of the University of Hong Kong (5). "There is an urgent need for multidisciplinary studies to explore the ecology of avian influenza viruses in wild birds and the environment to support ecological interpretation of the source of disease outbreaks in poultry."

Though my three examples all come from the United Kingdom, their implications are surely global. After decades of sometimes short-sighted reliance on antimicrobials and vaccines, the control of infectious diseases in the future will rest increasingly on ecological expertise, mathematical modeling, and a sensitive understanding of human and animal behavior.

References

1. **Hamer, W. H.** 1906. Epidemic disease in England. *Lancet* i:733–739.

2. **Hanratty, B., T. Holt, E. Duffell, W. Patterson, M. Ramsay, J. M. White, L. Jin, and P. Litton.** 2000. UK measles outbreak in non-immune anthroposophic communities: the implications for the elimination of measles from Europe. *Epidemiol. Infect.* **125:**377–383.

3. **Jansen, V. A., N. Stollenwerk, H. J. Jensen, M. E. Ramsay, W. J. Edmunds, and C. J. Rhodes.** 2003. Measles outbreaks in a population with declining vaccine uptake. *Science* **301:**804.

4. **Kermack, W. O., and A. G. McKendrick.** 1927. A contribution to the mathematical theory of epidemics. *Proc. R. Soc. A* **115:**700–721.

5. **Melville, D. S., and K. F. Shortridge.** 2006. Spread of H5N1 avian influenza virus: an ecological conundrum. *Lett. Appl. Microbiol.* **42:**435–437.

6. **Shirley, M. D., and S. P. Rushton.** 2005. Where diseases and networks collide: lessons to be learnt from a study of the 2001 foot-and-mouth disease epidemic. *Epidemiol. Infect.* **133:**1023–1032.

7. **Smith, K. F., A. P. Dobson, F. E. McKenzie, L. A. Real, D. L. Smith, and M. L. Wilson.** 2005. Ecological theory to enhance infectious disease control and public health policy. *Front. Ecol. Environ.* **3:**29–37.

23 Biocides in the Kitchen

Journals that require contributors to follow their conclusions with a further, prescient sentence under a rubric such as "Significance and Impact of the Study" often reveal authors struggling to say something profound. Gems I have collected over the years include "Our findings may assist understanding of the *in vivo* behavior of this organism" and "We have provided data others may use to explore further the growth characteristics we have identified here." I especially like that ambiguous word "may."

There was no such problem with the following assertion, which appeared at the top of a paper in the *Journal of Applied Microbiology* (1): "It refutes widely publicised, yet unsupported, hypotheses that use of antibacterial products facilitates the development of antibiotic resistance in bacteria from the home environment." The authors clearly saw their report as a demolition job on those who have attacked the current craze to incorporate potent antiseptics in a wide range of household products, from soaps, detergents, and lotions to impregnated cloths and chopping boards.

The attackers made two principal claims. First, the brandishing of biocides, such as triclosan, triclocarbon, and quaternary ammonium compounds, would quickly become self-defeating as organisms developed resistance to those substances. Secondly, and more importantly, cross-resistance meant that the practice would exacerbate the already considerable problems caused by drug-insensitive bacteria.

Were these hypothetical worries reflected in tangible problems in the real world? Maybe not if, as we once thought, the mechanisms by which some bacteria can withstand the actions of the two classes of agent were quite different. As we learn more about resistance to antiseptics, however, such reassurance looks less and less convincing.

Mutations in the *inhA* gene of *Mycobacterium smegmatis*, for example, confer insensitivity to both isoniazid and triclosan (5). Moreover, triclosan functions intracellularly as a site-directed enzyme inhibitor (3). Work in Stuart Levy's laboratory at Tufts University has shown that *Escherichia coli* mutants selected for resistance to pine oil disinfectant were, in addition, invulnerable to several antibiotics (6).

Studies in the United Kingdom on *Pseudomonas aeruginosa* at Unilever Research, Sharnbrook, and King's College London have also demonstrated that cross-resistance between antibiotics and biocides does occur, particularly in clinical strains (2). These authors suggested that alterations in permeability, as a consequence of the actions of aminoglycoside-modifying enzymes, might result in the acquisition of other forms of antimicrobial resistance.

The crusade against antibacterials in the kitchen has been led by Stuart Levy. His concerns include their capacity to foster potentially harmful changes in normal microbial populations and to add to the problems of antibiotic insensitivity. "Bacterial genes that confer resistance to antibacterials are sometimes carried on plasmids that also bear antibiotic resistance genes," he wrote in *Scientific American* (4). "By promoting the growth of bacteria bearing such plasmids, antibacterials may foster double resistance."

Levy observed that while routine kitchen cleansing is certainly important, conventional soaps and detergents reduce the numbers of potentially troublesome bacteria perfectly satisfactorily. "If we go overboard and try to establish a sterile environment, we will find ourselves cohabiting with bacteria that are highly resistant to antibacterials and, possibly, to antibiotics. ... It is not inconceivable that, with our excessive use of antibacterials and antibiotics, we will make our homes, like our hospitals, havens of ineradicable disease-producing bacteria."

So what of the paper that "refutes the widely publicised, yet unsupported, hypothesis" that vigorous, indiscriminate disinfection in the kitchen facilitates the development of insensitivity to antibiotics? It came from Eugene Cole and colleagues at two U.S. centers, DynCorp Health Research Services in Morrisville, NC, and Reckitt Benckiser Inc., Montvale, NY, in collaboration with Reckitt Benckiser in Hull, United Kingdom. They conducted an extremely thorough study in actual homes in North Carolina, New Jersey, and northeastern England.

For each of the regions, Cole and his collaborators placed newspaper advertisements to recruit 10 randomly selected homes where antibacterial products were used and 10 nonuser homes. The researchers screened applicants through telephone interviews to exclude any whose personal habits or histories could have compromised the results, for example, persons who had received antibiotics during the previous 30 days or who had been on long-term antibiotic therapy at any time during the year. All were questioned about their use of domestic biocides, which were recorded in inventories.

Samples were taken from sinks and other surfaces in the kitchen and bathroom and from soil likely to be tracked into the home. Bacteria targeted in these samples included coagulase-negative *Staphylococcus* spp., *Staphylococcus aureus*, *Pseudomonas* spp., *Acinetobacter* spp., and *Escherichia coli*. The investigators screened all 1,238 isolates for antibiotic sensitivity and tested selected resistant and sensitive isolates against triclosan, pine oil, parachlorometaxylenol, and quaternary ammonium compounds. MICs were determined.

The results (1) indicated that there was no significant or meaningful correlation between the antibiotic resistance patterns of any of the gram-positive or gram-negative potential human pathogens and their insensitivity to any of the four antimicrobial substances. In other words, the cross-resistance hypothesis examined in this very thorough study was not validated.

"In fact, the data for the 72 target bacteria isolates tested against the antibacterials show that for those isolates considered to have high MIC values against one or more of the active ingredients, most were fully susceptible to all of the preferred/alternative treatment drugs, and also were mostly found in non-user homes," the authors wrote. "Where high antibacterial MICs were found for some isolates, they were typically high for one, but not more than two of the active ingredients."

Just as interesting as these findings were the patterns of microbial life, and of antibiotic resistance, seen in the various locations. First, Cole and his coworkers recovered more target bacteria from homes not treated with antibacterial products than from those where they were in use. Second, all *S. aureus* isolates were sensitive to oxacillin and vancomycin, all *Enterococcus* isolates were sensitive to ampicillin and vancomycin, and all *Klebsiella pneumoniae* and *E. coli* isolates were sensitive to broad-spectrum cephalosporins. These findings were true for both user and nonuser homes.

Cole's results were telling, not least because they came from real homes rather than solely from bench research or work on clinical isolates. They certainly indicated that the contribution to antibiotic resistance from biocidal warfare in the kitchen may be considerably less than some commentators have alleged, but the findings will not terminate all disagreement over the robust commercialization of domestic biocides.

Few, if any, microbiologists would question the valid role that antiseptics have to play in combating food-borne disease. Cole and his team argued for "the targeted use of efficacious formulations of antibacterial products to combat a spectrum of emerging and often antibiotic resistant pathogens." However, even fewer can be happy to see potent antiseptics not only incorporated in a vast range of kitchen soaps, detergents, and liquids, but also impregnated in chopping boards, dish cloths, mattresses, toys, and high chairs. That is something upon which both sides of the argument may agree.

There is that word "may" again.

References

1. **Cole, E. C., R. M. Addison, J. R. Rubino, K. E. Leese, P. D. Dulaney, M. S. Newell, J. Wilkins, D. J. Gaber, T. Wineinger, and D. A. Criger.** 2003. Investigation of antibiotic and antibacterial agent cross-resistance in target bacteria from homes of antibacterial product users and nonusers. *J. Appl. Microbiol.* **95:**664–676.

2. **Lambert, R. J. W., J. Joynson, and B. Forbes.** 2001. The relationships and susceptibilities of some industrial, laboratory and clinical isolates of *Pseudomonas aeruginosa* to some antibiotics and biocides. *J. Appl. Microbiol.* **91:**972–984.

3. **Levy, C. W., A. Roujeinikova, S. Sedelnikova, P. J. Baker, A. R. Stuitje, A. R. Slabas, D. W. Rice, and J. B. Rafferty.** 1999. Molecular basis of triclosan activity. *Nature* **398:**383–384.

4. **Levy, S. B.** 1998. The challenge of antibiotic resistance. *Sci. Am.* **278**(3):46–53.

5. **McMurry, L. M., P. F. McDermott, and S. B. Levy.** 1999. Genetic evidence that InhA of *Mycobacterium smegmatis* is a target for triclosan. *Antimicrob. Agents Chemother.* **43:**711–713.

6. **Moken, M. C., L. M. McMurry, and S. B. Levy.** 1997. Selection of multiple-antibiotic-resistant (Mar) mutants of *Escherichia coli* by using the disinfectant pine oil: roles of the *mar* and *acrAB* loci. *Antimicrob. Agents Chemother.* **41:**2770–2772.

24 Conjectures and Realities

When our successors look back from well beyond the millennium, what will strike them most forcibly about our present-day relationship with the animalcules? Our fecklessness in misusing and thereby enfeebling antibiotics, our timorousness in developing environmental biotechnology, our innocence regarding the vast numbers of microorganisms yet to be discovered, or our precarious antimicrobial defenses, not only against hitherto-unrecognized pathogens, but also against novel recombinants of such familiar enemies as influenza virus?

These themes will no doubt provoke discussion, yet the outstanding issue may be something quite different—the extraordinary disparity in our attitudes toward two different constituencies of microbes with which we share the biosphere. On one hand, we are acutely preoccupied by possible risks in releasing genetically manipulated organisms (GMOs) into the environment and are spending large sums on evaluating and minimizing any such dangers. On the other hand, we do not fully understand day-to-day changes in the astronomically greater populations of animalcules which pose well-attested threats to human health and well-being.

Even today, let alone with the benefit of hindsight, the contrast is instructive. While fretting over conjectural problems which might, just conceivably, stem from scrupulously controlled experiments with harmless organisms, we do very little to monitor the evolution or population movements of a wide range of unambiguous pathogens whose capacity to disrupt human life is already amply documented. We behave like a theater manager who convenes a panel of experts to assess the remote possibility of one of next year's patrons disliking the color of the carpet while overlooking inadequate smoke detectors that put the whole enterprise at risk today.

It is, of course, entirely proper for conjectural ill consequences of releasing GMOs to be thoroughly addressed and then minimized or eliminated. This is an emerging technology, which ought to be applied with both technical and social prudence, and there are undoubtedly significant concerns, such as the possibility that bacteria deliberately released as cleanup agents may seize unforeseen opportunities to act as pathogens. However, the more carefully one examines such scenarios, the more it seems that the precision engineering of novel strains is our soundest strategy to avoid any such unforeseen hazards.

In regard to recombinant DNA work itself, the relaxing of guidelines in recent years indicates that we have been unduly cautious thus far. Can anyone doubt that the science of using GMOs for agricultural and environmental benefits is being applied more carefully and cautiously than any previous technology in the entire history of humankind?

Think of the requisite criteria. GMOs permitted to be released are minor variants of thoroughly familiar, exhaustively characterized organisms. Dedicated scrutiny has shown them to pose no threat to ourselves or other components of the biosphere, other than any specific targets against which they are consciously directed. They are individual strains, left to survive in the large, heterogeneous, well-adapted natural communities into which they are introduced. Their new traits are precise, discrete characteristics, incorporated for well-defined purposes. Their behavior postrelease is closely monitored, and they can be programmed to self-destruct after accomplishing their allotted tasks.

Now, as a midwinter exercise, contrast this scenario with that of the vast mixed microbial populations that constantly threaten our upper respiratory tracts, that have evolved as perpetual pathogens, and that possess formidable palettes of phenotypic and genotypic change. Medical microbiologists may wish it were otherwise, but we have only the scantiest knowledge of the behavior of these organisms and how they change from day to day, from month to month, and even from year to year.

True, networks such as the World Health Organization's influenza reference laboratories do sterling work. They provide invaluable information regarding new pandemic variants and have helped to generate our present understanding of "flu" virus evolution and of the importance of gene transfer between human and avian strains. Likewise, outbreaks of other acute respiratory infections inevitably attract the attention of public health authorities and are duly investigated and reported in the literature.

However, these are mere highlights of the ever-renewed attacks by viruses and bacteria on our respiratory mucosa. They become part of recorded knowledge because of the sheer severity of a disease, the large numbers of victims attacked, and sometimes the unusual features of the infection. In terms of both human misery and microbial colonization, a far greater fraction of respiratory illness goes unannounced.

As it happens, I am coughing energetically while writing this article, a result of a nasty infection "going around" London at the present time. My initial guess, when the first symptoms appeared several days ago, was that I had developed an adenoviral or streptococcal sore throat. Then, some of the features associated with respiratory syncytial virus (RSV) emerged. Now, my bronchial secretions tell me that bacteria have joined in the assault.

This self-diagnosis may or may not be correct. Either way (unless I progress to pleurisy or pneumonia over the next week and get admitted to the hospital), my affliction will fail to appear in any statistics on the nation's health. The organism(s) responsible will not be identified, and there will be no record concerning the size or spread of the outbreak. So, too, with the tens of thousands of similar infections in London this week and those in other population centers throughout the world this month, this year, and every year.

Most of us simply do not bother our physicians with our seasonal snuffles, so no throats are swabbed and no microsuspects are recovered. Pathogens are identified only when a sizeable proportion of individuals are sufficiently ill to warrant medical attention. Consider, for example, the annual epidemics of RSV infection, which have been increasingly recognized during the winter in recent decades, especially in large cities.

Animalcules: the Activities, Impacts, and Investigators of Microbes

Although RSV attacks people of all ages, most suffer in silence. Outbreaks become apparent largely through sudden rises in the numbers of infants admitted to hospitals with acute bronchiolitis. Even in the case of RSV, however, only the sketchiest of facts are recorded concerning the wider epidemics among adults.

Our ignorance is sharply illustrated by a survey of respiratory illness published in the *British Medical Journal* (1) by Karl Nicholson and colleagues at Leicester University. Introducing their paper, they made the following observation: "Remarkably little is known about the aetiology, prevalence and impact of acute respiratory infections among elderly people living in the community." This is an entirely valid assertion. It is also an astonishing one, illustrating a remarkable gap in our knowledge over 100 years after medical microbiology began.

Equally arresting is the conclusion of the Leicester survey: "Rhinoviruses are an important cause of debility and lower respiratory illness among elderly people. . . . The overall burden of respiratory infections in elderly people may approach that of influenza." Here is a crucial new discovery, with enormous implications for public health throughout the world. It could, and should, have been made many years ago.

What of gastrointestinal infections? It may be reassuring to believe that our surveillance over food- and waterborne maladies is much more extensive than our monitoring of respiratory pathogens. A comparison of the numbers of both types of outbreak recorded in the literature seems to support this view, yet the overall picture must be very similar. A tiny minority of investigations and surveys is largely eclipsed by countless bouts of diarrhea and vomiting in the great mass of humankind that never receive scientific scrutiny.

Meanwhile, committees ponder long and hard over the vanishingly small odds that fastidiously controlled experiments with intensively characterized nonpathogens will go awry. Strange, is it not?

Reference

1. **Nicholson, K. G., J. Kent, V. Hammersley, and E. Cancio.** 1996. Risk factors for lower respiratory complications of rhinovirus infections in elderly people living in the community: prospective cohort study. *Br. Med. J.* **313:**1119–1123.

25 Exterminating Pathogens

The name of Ali Maow Maalin deserves to be just as familiar to microbiology students as that of James Phipps, for it was Maalin, a 23-year-old Somalian cook, who in 1977 became the world's last naturally acquired case of smallpox, a disease rendered extinct by means of the form of vaccination first used by the English physician Edward Jenner on 8-year-old Phipps 2 centuries earlier. Some time over the next few years, hopefully, another name, that of the last person on Earth to contract paralytic polio caused by wild poliovirus, will join this historic duo.

But will poliomyelitis become the next communicable disease to be eradicated? Could rinderpest be the next candidate instead? Above all, what lessons can be learned from the eradication campaigns waged thus far, and are there any actual dangers in the extinction of pathogens?

Such was the agenda for a Dahlem workshop held in Berlin in 1997. It was one of a unique series of conferences, launched in 1974 but ending in 2006 as a result of economies imposed after the reunification of Germany. The published proceedings of these unique gatherings have been enormously influential as commentaries on evolving research fields and as free-ranging brainstorming sessions. Even today, the book from the 1997 conference, *The Eradication of Infectious Diseases* (1), stands as a major dossier on the realities of global and regional disease eradication efforts.

One consensus that emerged in Berlin was that there is no single microbial characteristic that guarantees either the success or failure of such campaigns. Even the most formidable obstacles, such as the existence of an animal reservoir, can sometimes be overcome. Witness the gradual elimination of malaria from Great Britain from 1850 onward as swamps were drained to exclude the mosquito vector. Another potent form of intervention is immunization, though the vaccine must prevent not just the disease, but infection, too.

The persistence of viable organisms in patients following recovery (especially if they cannot be detected easily) is one phenomenon that clearly can jeopardize wide-scale elimination. Thus, the possibility of eradicating tuberculosis, first postulated a century ago and enthusiastically revived with the advent of streptomycin in the 1950s, is compromised by the capacity of *Mycobacterium tuberculosis* to remain in healed lesions for many years. (Its presence is not revealed by a tuberculin test, which simply shows whether a person has been infected in the past.)

Speaking in Berlin, Frank Fenner of the John Curtin School of Medical Research in Canberra, Australia, argued that such persistence need not thwart the elimination of all infections. For example, the eradication of measles, one of two favored targets for early obliteration, can proceed satisfactorily despite the presence of the virus in the brains of people with subacute sclerosing panencephalitis. On the other hand, the occurrence of herpes zoster, whose victims may become infectious years after a chicken pox attack, calls for a "very long time frame" for any proposal to eradicate chicken pox.

One of the principal lessons to emerge from previous programs of disease extermination is the importance of thoroughly understanding the natural history of particular pathogens. For almost a year, from April 1927 to March 1928, no cases of yellow fever were reported anywhere in the Americas following successful control schemes initiated in Cuba 30 years earlier. Then, suddenly, new outbreaks occurred in Brazil and elsewhere, leading to the discovery that the virus was carried not only by *Aedes agypti*, but also by mosquitoes living in the forest canopy. If campaigners had recognized this reservoir earlier, they would probably not have wasted their time trying to combat the disease in certain areas.

During the World Health Organization's crusade against smallpox, seasonal declines in incidence offered a vulnerable point of attack.

Control measures implemented during these low points were much more effective than those adopted at other times. Although this feature of smallpox had not been fully appreciated at the outset, it was heavily exploited during the later stages of the campaign.

The World Health Organization's yaws campaign, one of its first initiatives in disease control, provides another lesson. Efforts over 2 decades up to the late 1960s reduced prevalence levels below 0.5% in many African countries. However, yaws reappeared when the program was prematurely discontinued so that personnel working in mobile field units could be redeployed in primary health care.

Discussions in Berlin indicated that while dedicated eradication programs can still draw money and skills away from other health services, they have generally strengthened the global medical infrastructure. They have also focused international attention on the need to ensure that all populations have access to basic medicine.

Participants felt, too, that civil unrest and wars did not pose insurmountable barriers to disease control. "Days of tranquility," when combatants agree to a cease-fire to allow activities such as polio vaccination, may even contribute to a more lasting peace in troubled regions.

Eliminating pathogenic microorganisms is not cheap, and global initiatives must be adopted even in places where a particular infection does not pose a serious problem. Still, the rewards are immense. Polio eradication would mean a global financial saving of around $1.5 billion on vaccine alone, and while costs are incurred over a limited period of time, the many benefits of disease elimination accrue in perpetuity. Only economists are discomfited by a scenario that defies their traditional methods of analysis.

Are there any drawbacks to disease extermination? One touched upon in Berlin was a possible conflict with the post-Rio de Janeiro consensus on biodiversity. To some degree, this issue can be sidestepped by distinguishing between absolute extinction (as with smallpox) and eradication, defined as the permanent reduction of worldwide infection to zero.

Neonatal tetanus, for example, could be totally eliminated by immunization of pregnant women and by aseptic care of the umbilical cord, but this would leave *Clostridium tetani* spores in the soil and occasional cases of tetanus in adults. The more radical step of exterminating pathogens entirely is bound to pose sharper dilemmas in relation to biodiversity as others become candidates for extinction and as we learn more about the fine texture of the biosphere.

A related question is whether the removal of pathogens may create ecological niches to be filled by existing or genetically altered replacements. Most simplistically, the human cell receptors hitherto used almost exclusively by polioviruses might be considered to be "vacated" following polio eradication. Could other viruses evolve to occupy the same sites?

There is no compelling evidence to support the niche theory. On the other hand, it is by no means crazy to speculate that the existence of smallpox (and vaccinia) virus has prevented human infections with closely related poxviruses of other animals. Monkeypox in humans, which spread in 1997 through war-torn Zaire, became apparent for the first time only after the eradication of smallpox.

More broadly (and despite the surprisingly buoyant assertion by the World Health Organization in May 2008 that "the global burden of disease is shifting from infectious diseases to noncommunicable diseases), recent decades have certainly seen a dramatic rise in new and newly recognized infections. There are credible explanations for most of these developments. *Listeria*, for example, has become more important as an agent of food poisoning with the greater use of refrigeration at temperatures satisfactory for the organism to proliferate. Nevertheless, the emergence of novel infectious diseases could have links with the control and elimination of old ones.

Aside from the "hygiene hypothesis" that links cleaner conditions in childhood with inadequate challenge to the developing immune system and thus greater predisposition to allergy later in life, is it perhaps time for a systematic study of this intriguing possibility?

Reference

1. **Dowdle, W. R., and D. R. Hopkins (ed.).** 1998. *The Eradication of Infectious Diseases.* Wiley, New York, NY.

26 Learning from Denmark

Since the beginning of the new millennium, Denmark's pig producers have entirely relinquished the use of antibiotics as growth promoters. They took this unprecedented step as a voluntary response to increasing professional and public concern about the proliferation of drug-insensitive bacteria, such as salmonellae, and their spread from farm animals to humans.

The Danish move is the most recent in a series of initiatives that have placed the northern European countries in the forefront of the worldwide battle against antibiotic resistance. In 1997, for example, Finnish microbiologists announced their success in halving the rate of recovery of erythromycin-insensitive group A streptococci from throat swabs and pus samples. Helena Seppälä and her colleagues (3) attributed this change to stringent new national guidelines on the prescribing of macrolides instituted 4 years previously.

The contributions made by medical applications and misapplications of antibiotics to the burgeoning global pool of drug resistance are probably far greater than those attributable to animal husbandry. Nevertheless, the practical consequences in terms of untreatable infections are so serious that any significant diminution is welcome.

Animalcules: the Activities, Impacts, and Investigators of Microbes

The initiative of 2000 in Denmark was the latest of several steps taken voluntarily by that country's pig and poultry farmers to curb the mass dissemination of antibiotics to promote growth. The government, too, has imposed increasingly strong controls over veterinary products.

In 1995, it ruled that these products would in future be available only from pharmacists. Veterinarians could no longer sell them for profit (except in emergencies). The removal of their financial interest in what they prescribe, and in what quantities, has been a key policy limiting the use of antibiotics in Denmark. Since the restriction of these and other drugs to pharmacies, the quantity prescribed to combat animal disease has fallen significantly.

Nineteen ninety-five likewise saw the creation of a nationwide program to tackle a serious problem of *Salmonella* species in pig meat. Particular attention focused on the strain of *Salmonella enterica* serotype Typhimurium known as definitive phage type 104 (DT104). It is usually insensitive to ampicillin, chloramphenicol, streptomycin, sulfonamides, and tetracyclines, and an increasing proportion of isolates are also relatively resistant to fluoroquinolones. If meat inspection staff find one pig that is infected with this organism, the entire herd has to be slaughtered. Inspectors test the animals in a random but intensive screening program at abattoirs.

It was also in 1995 that researchers at the Danish Veterinary Laboratory in Copenhagen published evidence that avoparcin, then included in pig feed as a growth promoter, could cause resistance and cross-resistance to vancomycin. Against a feverish background of parliamentary and public opposition to the very idea of giving antibiotics to "healthy animals," the government immediately banned avoparcin from pig fodder.

Contrary to some farmers' fears, the removal of all growth-promoting antimicrobials from pig food (which was actually in progress before the 1 January 2000 deadline) was accomplished without any serious ill effects. The only significant repercussion was that the animals tended to produce slightly looser dung, and in some cases, they developed diarrhea. However, these changes (attributable to shifts in the intestinal flora following the removal of antibiotics) usually abated within a month and otherwise could be prevented by dietary modification.

More importantly, these alterations in feedstuffs plus those necessitated by the termination of artificial growth promotion did not load farmers with greatly increased costs. The additional annual outlay was less than $0.65 per pig.

It is now over 49 years since the British bacteriologist E. S. Anderson, a pioneer of microbial drug resistance, called in *Nature* (1) for "a reexamination of the whole question of the use of antibiotics and other drugs in the rearing of livestock." His concern was based on meticulous research in which he traced the emergence of ampicillin-resistant *S. enterica* serovar Typhimurium strains to the use of the agent to treat and/or prevent infections in calves.

Working for the then Public Health Laboratory Service in London, Anderson presented evidence that a particular phage type of the organism had been distributed around Britain by livestock movements and had caused human infections. He argued that if farmers improved their standards of animal husbandry, they could stop using antimicrobial drugs for the twin purposes of disease prophylaxis and growth promotion.

Anderson's work led to a government inquiry 4 years later and to restrictions on the incorporation of tetracyclines and penicillins into feedstuffs. However, opposition to controls of this sort has continued, both in the United Kingdom and elsewhere, to the present day. Critics, questioning the contribution of intensive agriculture to bacterial resistance, have argued that it would be financially prohibitive for farmers to stop dousing entire herds and flocks with antibiotics.

The Danish experience is showing that Anderson was right. "The standard of animal husbandry management in Denmark has been the critical factor in removing antibiotic growth promoters," wrote Debra Robertson-Welsh and Verner Wheelock (2) in their report *Food Safety and Pig Production in Denmark*. "Moreover, there has been no corresponding increase in the therapeutic or prophylactic use of antibiotics during or since the removal of growth promoters from feed."

Of course, the situation in Denmark requires continued monitoring. One purpose of the surveillance now in place is to ensure that farmers do not simply obtain antimicrobials from another source (as was widely rumored after the curbs on penicillins and tetracyclines in the United Kingdom 3 decades ago). The second aim is to gather evidence on the consequences of the voluntary ban for the pattern of resistance among enterobacteria in farm animals and in the community at large.

Given Denmark's preeminence as a bacon exporter, particularly within Europe, the new arrangements bring benefits for both public health and public relations. It is strange that many food campaigners are less interested in concrete measures of this sort, adopted to combat the considerable, proven menace of drug resistance, than in fostering hysteria over remote conjectural hazards of genetically modified food.

References

1. **Anderson, E. S., and M. J. Lewis.** 1965. Drug resistance and its transfer in *Salmonella typhimurium*. *Nature* **206:**579–583.

2. **Robertson-Welsh, D., and V. Wheelock.** 2000. *Food Safety and Pig Production in Denmark.* Verner Wheelock Associates, Skipton, United Kingdom.

3. **Seppälä, H., T. Klaukka, J. Vuopio-Varkila, A. Muotiala, H. Helenius, K. Lager, and P. Huovinen.** 1997. The effect of changes in the consumption of macrolide antibiotics on erythromycin resistance in group A streptococci in Finland. *N. Engl. J. Med.* **337:**441–446.

27 Protozoa and
Lurking Pathogens

I can still recall my confusion, as a student 50 years ago, over the evolution of microorganisms. Many textbooks portrayed the first animalcules as simple heterotrophs, wholly dependent for their energy and materials on the rich chemistry of the primeval soup. Immeasurably slowly, they acquired the capacity to synthesize enzymes, and thus essential substances, as these were successively depleted in the environment. It was a persuasive picture of the stepwise emergence of metabolic pathways.

However, other books seemed to show a more arid world in which the original organisms were autotrophs, photosynthesizing or transforming chemicals to acquire their energy and to build up complex materials from carbon dioxide. Opportunities for living as parasites or saprophytes lay well into the future, since multicellular creatures had yet to evolve.

Part of the difficulty with questions of this sort is the conceptual challenge of imagining what we now call the biosphere before it was the biosphere. How, then, to respond to the suggestion by Michael Brown of Aston University, Birmingham, United Kingdom, that billions of years of evolution produced pathogens long before there were animals for them to infect? Given that pathogens are defined by reference to their hosts, this sounds rather like archaeologists finding evidence of keys in an ancient civilization that had not yet invented locks.

Animalcules: the Activities, Impacts, and Investigators of Microbes

Still, there is telling support for Brown's suggestion, based on the ubiquity of protozoa in the present-day world. He believes that, while protozoal grazing of bacteria is well recognized, we have neglected a very different scenario in other environments. Here, bacteria invade protozoa, replicate, and emerge better fitted to enter other protozoa, as well as more resistant to antibiotics, biocides, and other stresses.

One can envisage the two forms of microbial life coexisting and coevolving before we and other animals appeared. The principal selective factors affecting bacteria were predation by protozoa and environmental stresses, such as drying, freezing, and oxidation. The organisms that endured were those that were able to cope with both and were well equipped for survival as putative pathogens.

A major plank in Brown's hypothesis is the significance of protozoa today as reservoirs of animal (including human) pathogens. The best-known example is, of course, *Legionella pneumophila*. It not only thrives in amoebae, but also, as shown by Stanley Falkow and colleagues (3), emerges with a greatly enhanced ability to invade macrophages and epithelial cells. Likewise, the growth of *Mycobacterium avium* in amoebae increases not only its capacity to enter macrophages, but also its virulence (4).

Brown cites two examples of the possible (but possibly huge) medical relevance of these discoveries. First, badgers have been culled in various parts of Britain in recent years in experiments designed to clarify their role as reservoirs of *Mycobacterium bovis*, which can infect cattle, but how important as sources of infection are protozoa, which are widespread in the environment, including sewage and silage?

Secondly, what of their role in harboring *Escherichia coli* O157 strains, which have caused major outbreaks of hemolytic-uremic syndrome in Britain and North America in recent years? Brown and his coworkers have demonstrated (1) that this organism, too, not only survives inside *Acanthamoeba polyphaga*, but also replicates in this environmental protozoon, which occurs in many soils and effluents.

"Our results indicate that protozoa have important potential to act as vehicles for the dissemination of *E. coli* (and possibly other pathogens) in the environment," Brown says. He points out that amoeba trophozoites could offer a protective niche for *E. coli* against adverse conditions. This would be especially true if the organism can survive within amoeba cysts, as has been demonstrated for *L. pneumophila* and *Vibrio cholerae.*

"Bacteria trapped within amoeba cysts could be blown through the air and enhance distribution. Grazing cattle almost certainly ingest protozoa in silage and grass, so if the protozoa contain pathogens, this could be a significant route of infection between and within cattle herds."

The link proposed by Michael Brown between bacterial evolution in the environment and pathogenicity is the general stress response (GSR). Building on evidence of its dual role in survival and virulence, he argues that the coevolution of bacteria and protozoa equipped some species of bacteria both for persistence in the environment and for invasion of, and survival in, animal cells. They could, for example, have developed resistance to phagocytosis and serum killing even before animals with these defensive routines had evolved.

An *rpoS*-encoded sigma factor is the master regulator in a complex network of stationary-phase-responsive genes involved in the GSR in *E. coli.* Brown points out that, in addition to their role in resistance/survival, *rpoS* activity and the subsequent GSR response are associated with changes in pathogenicity. Thus *rpoS* regulates genes that are important for the virulence of *Salmonella enterica* serovar Typhimurium, promoting its aggressiveness in invading Peyer's patches in the gut.

Future research findings may confirm or refute Brown's conjectures about the past. Of more immediate practical importance is the evidence suggesting that protozoa, functioning as Trojan horses of the microbial world, may play a far greater role than has yet been recognized in harboring and disseminating pathogens. Much of this activity is likely to escape detection by conventional laboratory culture techniques.

Commenting on the apparent decline in *Helicobacter pylori* infection in developed countries, Martin Blaser (2) has argued that "to maintain its niche in the human biosphere, *H. pylori* must be transmitted from person to person." In light of Michael Brown's findings and speculations, I wonder whether such confidence is fully justified for this or many other pathogens.

References

1. **Barker, J., T. J. Humphrey, and M. W. Brown.** 1999. Survival of *Escherichia coli* O157 in a soil protozoan: implications for disease. *FEMS Microbiol. Lett.* **173:**291–295.

2. **Blaser, M. J.** 1999. Where does *Helicobacter pylori* come from and why is it going away? *JAMA* **282:**2260–2262.

3. **Cirillo, J. D., S. Falkow, and L. S. Tompkins.** 1994. Growth of *Legionella pneumophila* in *Acanthamoeba castellanii* enhances invasion. *Infect. Immun.* **62:**3254–3261.

4. **Cirillo, J. D., S. Falkow, L. S. Tompkins, and L. E. Bermudez.** 1997. Interaction of *Mycobacterium avium* with environmental amoebae enhances virulence. *Infect. Immun.* **65:**3759–3767.

28 What Is Virulence?

A dialysis bag is an odd but appropriate artifact to symbolize the interface between two investigational techniques. Described during a meeting at the Royal Society in London, this one sits at the boundary between the chemical approach to microbial pathogenicity and the analysis of virulence through molecular genetics. While the former is already a mature discipline, the latter is in its initial, vigorous phase of growth.

The devisers of the dialysis bag were Carlos Orihuela and colleagues at the University of Texas Medical Branch, Galveston, TX, and Information Dynamics Inc. in Nassau Bay, TX. They use it to explore the difference between the behaviors of *Streptococcus pneumoniae* growing in vitro and in vivo. Inoculated with pneumococci, the bag is surgically implanted in the abdominal cavity of a mouse. Several hours later, it is recovered, and the bacterial products are evaluated.

In the study presented in London, the Texas group showed by sodium dodecyl sulfate-polyacrylamide gel electrophoresis and other techniques that production of several different proteins was enhanced significantly in vivo. The organisms also adhered 10 times more avidly to lung epithelial cells than those cultured in vitro. Analysis of virulence gene expression by Northern dot blotting revealed that transcription of the genes for pneumolysin and for synthesis of type 3 capsule more than doubled.

Animalcules: the Activities, Impacts, and Investigators of Microbes

No doubt Harry Smith would have relished the opportunity to harness techniques of this sort when he pioneered the chemical study of pathogens some 40 years ago. The man who discovered why *Brucella abortus* preferentially attacks the fetal membranes of cattle, sheep, and goats (because they are rich in erythritol), Harry is now an emeritus professor at Birmingham University, Birmingham, United Kingdom. His continuing fascination with the specific determinants of microbial disease was attested by his role as coorganizer of the Royal Society meeting.

Our comprehension of microbial disease is, of course, now being revolutionized by the sequencing of the genomes of successive pathogens. However, this is by no means the only development that is transforming the picture. Equally important is the emergence of powerful methods for identifying virulence genes and their expression in vivo. Third, knowledge is burgeoning regarding the signals associated with the adhesion of bacteria to target cells, invasion, and interference with host cell functions.

While a combination of bench and animal studies led to the identification of nutritional factors, such as erythritol, in contagious abortion in cattle, other diseases have proved considerably more difficult to fathom. This is because many crucial aspects of communicable disease are not manifest in organisms cultured in laboratory glassware—even those freshly isolated from diseased tissues.

Screening with tissue-cultured host cells is superior in this sense to lifeless culture media. Nevertheless, it fails to reveal all virulence genes because such cells reflect neither the sophisticated architecture of the target organ nor the complex interactions between pathogens and host immune defenses.

The new techniques, however, allow experimenters to detect hitherto-unrecognized pathogen genes and their expression during the course of an infection. They can, for example, go far beyond our knowledge of constitutive gene products to reveal genes that are expressed in vivo in an organ- or cell-specific fashion.

As Andrew Camilli of Tufts University explained, in vivo expression technology has been developed with a variety of reporter systems, permitting either in vivo selection or ex vivo screening. "Selectable gene fusion systems generally allow for the complementation of a bacterial metabolic defect that is lethal *in vivo*, or for antibiotic resistance during the course of *in vivo* antibiotic challenge.

"In contrast, the screenable gene fusion system uses a site-specific DNA recombinase which, when expressed *in vivo*, excises a selectable gene cassette from the bacterial chromosome. Loss of this cassette can then be either screened or selected for *ex vivo*."

Investigators are using the recombinase-based in vivo expression technology system to detect genes that are transcriptionally induced during an infection. It is sufficiently powerful to reveal those that are expressed at low levels and those that function only transiently. Recombinase-based in vivo expression technology helps to demonstrate the spatial and temporal patterns of gene expression as an infection progresses.

Signature-tagged mutagenesis (STM), another novel methodology, allows investigators to test many different insertional mutants simultaneously in a living host, which selects against strains whose virulence has been attenuated. It was initially validated using *Salmonella* and a mouse model of typhoid fever by the isolation of many genes now known to be important for virulence.

Brian Sheehan and his colleagues have applied STM to study the highly contagious, rapidly fatal pleuropneumonia caused by *Actinobacillus pleuropneumoniae* in pigs. Working at Imperial College School of Medicine in London and the Royal Veterinary College, Hatfield, United Kingdom, they first screened 48 sequence-tagged mini-Tn*10* mutants by intratracheal inoculation.

This led them to identify four attenuated mutants and to characterize the DNA flanking each transposon insertion site. In one, the transposon had inserted in a gene not resembling any to be found in the public databases. The others contained insertions in loci encoding an ABC-type transporter, an NAD(P) transhydrogenase, and TonB, which had previously been implicated in the virulence of several different organisms.

Sheehan and his coworkers pointed out that TonB *Salmonella enterica* serovar Typhimurium mutants were attenuated when introduced by the mucosal route, but not when given intraperitoneally. "TonB was not identified during the STM analysis of *S. typhimurium*, emphasising the importance of the route of infection in identifying genuine modulators of bacterial virulence," they concluded. "STM holds promise for identifying *A. pleuropneumoniae* genes for natural porcine infection, which may be exploited in designing strategies to prevent the disease."

STM is clarifying the relationships between salmonellae and their target species, too. Tim Wallis and colleagues at the Institute of Animal Health, Compton, United Kingdom, have used it to identify two *S. enterica* serovar Dublin mutants with apparently altered host-specific phenotypes.

One, which seemed to be attenuated only in mice, contained an insertion in the *sseD* gene in *Salmonella* pathogenicity island 2 (SPI-2). This locus (SPI-2) was previously associated with the systemic survival of *S. enterica* serovar Typhimurium in mice. The other mutant, attenuated only in calves, had the insertion upstream of the *Vibrio cholerae* gene *hcp*, which had not previously been identified in salmonellae.

Wallis and collaborators believe that both genes code for secreted proteins. In contrast to their initial suspicions, they discovered that the *sseD* mutant was avirulent in both mice and calves. They believe their work provides the first evidence of the importance of SPI-2 as regards systemic salmonellosis in calves.

The Compton team also used in vivo tests to confirm that the putative *hcp* mutant was attenuated in calves but remained virulent in mice. This, in turn, is the first piece of work showing that the *hcp* locus influences virulence and potentially serotype-host specificity.

Techniques and advances of this sort are indeed impressive, yet as Richard Moxon of Oxford University pointed out during the Royal Society meeting, they do not solve the entire problem of how microbes cause disease. Practical and semantic difficulties remain in distinguishing so-called virulence factors from the many determinants promoting general fitness. "Despite the awesome power of molecular biology," Moxon said, "we are still struggling to come to grips with the simple question of 'what is a pathogen?'"

As in many other sectors of biology, we need to remain vigilant over the seductions of reductionism and determinism.

III. The Human Context

29 Questionable Experiments

Imagine that you understand little or nothing about microorganisms, genetic modification, toxins in the natural world, or the biological control of plant pests. Now consider the following. A scientist tells you that he has been breeding venomous scorpion spiders in his laboratory. He's also been cultivating some extremely infectious microbes, and he plans to alter them genetically so that they produce the same potent poison as the scorpion. The poison is called a neurotoxin because it attacks nerve cells, causing paralysis and death. The scientist now plans to produce astronomical numbers of his mutant microbes and release them in the countryside.

How do you respond to what you have heard? I believe most nonscientists would react with feelings somewhere along a scale ranging from acute alarm to anger and hostility. That is why the scorpion toxin scenario, which is not science fiction but reality, is a peculiarly stark example of a challenge inherent in several areas of modern bioscience. How should scientists conduct themselves when they wish to perform potentially controversial experiments, experiments they believe are amply justified and entirely safe yet which can sound dangerous, reckless, and irresponsible.

In December 2007, two researchers did publish a paper on this type of work. They were Chengshu Wang of the Shanghai Institutes for Biological Science in Shanghai, China, and Raymond St Leger of the University of Maryland. Their paper in *Nature Biotechnology* (3) is entitled "A scorpion neurotoxin increases the potency of a fungal insecticide." It demonstrates that the expression of an insect-specific scorpion toxin in *Metarhizium anisopliae* heightens the toxicity of this fungus 22-fold against tobacco hornworm caterpillars and 9-fold against yellow fever mosquitoes.

"Despite intensive efforts, biocontrol agents have not met expectations, primarily because of their poor efficacy and inability to compete with cheaper chemical peptides," the authors write. "Developing host-specific strains based on our hypervirulent pathogenic fungus, which incorporate measures to avoid environmental contamination, should open the way for cost-effective control of a range of insect pests of agricultural and medical importance."

The calm reception of Wang and St Leger's important research may be contrasted with that accorded to some similar experiments and field trials carried out in the United Kingdom in the late 1980s and early 1990s. The leader of this work was David Bishop at the Institute of Virology in Oxford, of which he was then director. His intention was to use a baculovirus carrying a scorpion toxin gene to attack pests, such as the pine beauty moth, which causes considerable damage to pine forests in Scotland.

It is the caterpillars that are the culprits, capable of destroying many hectares of lodgepole pine trees. They can be combated to some degree by spraying chemical pesticides, such as Dimilin. However, an environmentally far more desirable solution could be to use a natural enemy, and the one chosen by David Bishop was a nuclear polyhedrosis virus, *Autographa californica*. Registered as an insecticide in the United States, this baculovirus does not infect or harm vertebrates, plants, or invertebrates. Its only targets are a small number of pestiferous moths.

When Bishop initiated his program of research, there was already public agitation in the United Kingdom over the real and (mostly) imagined hazards of releasing genetically modified organisms into the environment. Some sections of the media were reflecting and amplifying these concerns, yet David behaved in an exemplary fashion, proceeding exceptionally cautiously, both technically and socially, toward his goal.

He showed scientific prudence by first developing a labeled virus, in order to demonstrate that he could monitor its movement and distribution if it was released into the environment. As described in *Nature* (1), he introduced a noncoding synthetic oligonucleotide sequence into an intergenic region of the genome at a position that deletion experiments had shown to be nonregulatory and expendable. After being extensively tested in the laboratory with regard to its genetic stability and biological phenotype, and in comparison with unmarked virus, the modified version was then released in a "cabbage patch" ecosystem. This experiment, the first to be approved by the United Kingdom's newly established Advisory Committee on Genetic Manipulation, confirmed that the virus could be sensitively traced after release.

Only after completing these and other related studies did the Oxford group conduct its first field trial with a baculovirus containing a scorpion toxin gene. Again supported by extensive laboratory assays, the modified virus was tested at the university's field station on plots of cabbages infected with the cabbage looper caterpillar. As reported in *Nature* (2), the modified virus killed faster, resulting in less crop damage, and also appeared to reduce the secondary cycle of infection compared with the wild-type virus.

Meanwhile, David Bishop was doing everything imaginable to discuss his work openly and to listen to and respond to the concerns of people in and around Oxford. He made a video explaining why he believed his approach to biological control was both necessary and safe. He did countless media interviews, called public meetings, and posted notices in his local newspaper to explain what he was doing. He was willing to travel anywhere, any time, to face all manner of audiences, including hostile ones.

Paradoxically, however, it seemed that these proactive initiatives did not entirely reassure the Oxford public. If anything, they served to heighten public anxiety. What happened next is open to a variety of explanations. David Bishop was sacked from his post. The Institute of

Virology's funding body, the Natural Environment Research Council, had decided that "the institute's mission had changed and that Bishop was no longer the right person to be in charge."

It is possible that the furor in Oxford over the baculovirus work, amply reflected in national media coverage, had nothing to do with Bishop's demise. That was the Natural Environment Research Council's position. David's only comment, in a three-page statement, was, "If one pioneers work with a virus insecticide, genetically engineers it, puts in an insect-selective toxin gene, one that comes from a scorpion, and then does field trials with this improved insecticide, then one is fair game for the press."

Whatever the whole truth may be, the Oxford episode does serve to illustrate an uneasy question about scientific work. Are there potentially rewarding experiments that should be considered but then rejected, not because they are hazardous, but because nonspecialists may feel them to be hazardous? Remember, "nonspecialists" here can well include scientists in disparate disciplines. The leader of the Oxford protests was a senior lecturer in the university's department of metallurgy.

From the very beginning of genetic modification, there have been understandable concerns over potentially dangerous experiments. An early furor surrounded the question of whether polyomavirus genes should be inserted into *Escherichia coli* to determine whether they and their oncogenic potential might be transferred from gut bacteria into human tissues. Much more recently, questions were asked about the risks inherent in recombining H5N1 and human influenza viruses to see how hazardous the hybrid might be.

However, these are issues, like the dilemma of whether the world's remaining stocks of smallpox virus should be destroyed, upon which even specialists have been divided. Far more ticklish are the conundrums that arise when there is expert unanimity that certain procedures pose an acceptably low risk, or no risk at all, while the words used to describe them conjure up images that other people find unacceptable and threatening. I do not know the answer to this question, but I do believe it is likely to come up in other guises prompted by advances in molecular genetics, especially, perhaps, the molecular genetics of microorganisms.

References

1. **Bishop, D. H. L.** 1986. UK release of genetically marked virus. *Nature* **323**:496.

2. **Cory, J. S., M. L. Hirst, T. Williams, R. S. Hails, D. Goulson, B. M. Green, T. M. Carty, R. D. Possee, P. J. Cayley, and D. H. L. Bishop.** 1994. Field trial of a genetically improved baculovirus insecticide. *Nature* **370**:138–140.

3. **Wang, C., and R. J. St Leger.** 2007. A scorpion neurotoxin increases the potency of a fungal insecticide. *Nat. Biotechnol.* **25**:1455–1456.

30 Lyme Disease:
The Public Dimension

During a recent talk, I mentioned the increasing number of reports in medical journals about people who have harmed themselves by following health advice, and sometimes buying weird nostrums, from the Internet. Recent cases include a man who developed severe nephropathy after consuming a Chinese herb "to enhance his liver" and another whose online purchase of prednisolone for a self-diagnosed illness triggered the development of bilateral cataracts and glaucoma.

"But those examples are not typical," said someone in the audience. "You are overlooking the enormous amount of authoritative guidance and real help now available on the Net. People are learning about rare syndromes and new therapies, bringing important information to the attention of their doctors, and securing earlier, more appropriate treatment, as well as saving time and money. All of this is revolutionizing medical care. There are bound to be a few difficulties, but they are totally outweighed by the benefits."

My critic had a point, yet I remain concerned about the huge quantity of nonsense that now flows through the global electronic highway and its practical consequences. This is certainly an area ripe for a dedicated piece of research. Notwithstanding some real gains, I suspect that a systematic study of health advice available through the uniquely powerful, unregulated Internet would reveal far more disquieting examples than beneficent ones.

Meanwhile, we can learn from specific case histories about the impact of the Internet on medical research and practice. An especially telling example is Lyme disease. The story of this condition, since it hit the headlines just over 30 years ago, is instructive because it vividly illustrates both the positive importance and the drawbacks of greater public involvement in the advancement of medical science and health care. On one hand, concerned citizens, rather than medical experts, first highlighted the full significance of Lyme disease. On the other hand, public clamor, much of it centered around the Internet, has now derailed the introduction of a highly effective vaccine against the infection.

The story began in 1975 when a vigilant mother in Connecticut noticed that 12 children in the village of Old Lyme (population, 5,000) had developed an illness that had been diagnosed as juvenile rheumatoid arthritis. Local doctors appeared unconcerned, so the woman, increasingly perplexed and worried, reported the matter to her state health department. Around the same time, another villager independently telephoned the Rheumatology Clinic at Yale University to describe an "epidemic of arthritis" in her family. This pattern, too, had not been picked up by the otherwise meticulous health surveillance machinery of the state of Connecticut.

At first, officials were deeply skeptical about the claims from Old Lyme and impatient with the villagers' demands that the mystery be investigated. Who had ever heard of arthritis appearing as an epidemic? Arthritis was not an infectious disease, it was a degenerative condition associated with ageing. There was simply no way in which it could spread around a community like chicken pox or measles.

Fortunately, researchers at Yale did take the women seriously and began to monitor what was happening. By 1977, they were convinced that there was indeed an outbreak of arthritis around Old Lyme. As well as aching joints and a stiff neck, the disease caused headache and fever. Two other striking characteristics were that it tended to begin in the summer and appeared among children or adults several weeks after an unusual sort of spot had suddenly developed on the skin.

The first real clue to the mystery came when one patient recalled having been bitten by a tick at the site of the spot. Investigators discovered that a particular type of tick, usually carried by deer, was the carrier of an organism that might cause the disease. Further detective work led to the isolation of a characteristic spirochete from the tick and the demonstration that there were antibodies against it in the blood of Lyme arthritis patients. Finally, Willy Burgdorfer and colleagues described the organism, isolated from patients' blood and now called *Borrelia burgdorferi*, in the *New England Journal of Medicine* (2).

So far, so good. Activists had helped to focus professional attention on a condition that was in danger of being overlooked. Then, in the years following, a bandwagon began to roll as campaigning efforts attracted tens of millions of research dollars annually. Lobbyists portrayed Lyme disease as a national plague and wielded considerable influence on the research agenda. For microbiologists who believed the condition was little more than a nuisance in certain geographically restricted areas and who emphasized that it was treatable with antibiotics, the threat was being grossly exaggerated. Just as regrettable was the distortion of scientific research by demands from activists.

"These lay pressure groups are interfering with research. . . . There is science, and there is nonscience, and nonscience doesn't belong at a scientific meeting," said Durland Fish of New York Medical College in Valhalla. He was speaking as a member of the program committee for a conference on Lyme borreliosis held in Arlington, VA, in 1992. Fish's irritation followed the reinstatement of several papers originally rejected by the committee as not being of the requisite standard. Written by nonacademic clinicians, the papers were put back on the agenda largely as a result of pressure from patient support groups. The principal issues at stake included the question of whether the patients described in the controversial reports were really suffering from Lyme disease, which was being overdiagnosed, and whether they were receiving valid therapies.

One might imagine that the development of a recombinant Lyme disease vaccine, and its approval by the FDA in 1998, would have been wholly welcome to the campaigners and would have ended their by now overheated campaigning. What happened instead, however, was that the activists who had urged the development of a vaccine greeted its arrival with hostility. As described by Lise Nigrovic and Kimberly Thompson of the Children's Hospital Boston and Harvard University, Boston, MA (1), their angry response, linked with adverse media cov-

Animalcules: the Activities, Impacts, and Investigators of Microbes

erage, fears of side effects, and declining sales, led the manufacturer to withdraw the product.

According to Edward McSweegan, program officer for Lyme disease research at the National Institute of Allergy and Infectious Diseases in the early 1990s, licensing of the vaccine confronted campaigners with the problem of "how to sustain public anxiety (and donations), media attention and political clout against the evidence-based reality of a bacterial infection that was antibiotic-responsive, non-fatal, non-communicable, geographically focused and preventable through vaccination."

The activists' answer was to target the vaccine's imperfect efficacy, projected cost, and potential booster requirements. "*Ad hominem* attacks on individuals involved in the vaccine trials quickly followed, stoked by simmering animosity between many patient activists and clinicians over the appropriate diagnosis and treatment of Lyme disease. These personal attacks—and anecdotal stories about Lyme disease in general and the vaccine in particular—took place on the internet."

Campaigners' websites also purveyed "misleading information about the vaccine, personal 'vaccine victims' stories, and newsgroup bulletin boards offering a repetitive stream of misinformation, libel and quack treatments." McSweegan argues that, because this was barely answered by researchers, vaccine manufacturers, or public health officials, "the public opinion battles . . . were fought, and lost, in cyberspace."

I fervently hope that lessons have been learned on all sides from this initially positive but eventually dismal story.

References

1. **Nigrovic, L. E., and K. M. Thompson.** 2007. The Lyme vaccine: a cautionary tale. *Epidemiol. Infect.* **135**:1–8.

2. **Steere, A. C., R. L. Grodzicki, A. N. Kornblatt, J. E. Craft, A. G. Barbour, W. Burgdorfer, G. P. Schmid, E. Johnson, and S. E. Malawista.** 1983. The spirochetal etiology of Lyme disease. *N. Engl. J. Med.* **308**:733–740.

31 Blatant Opportunism

Every human generation has to recognize anew, and if possible anticipate, the relentless opportunism of bacteria, viruses, and other animalcules. Whatever changes occur in the way we live our lives or modify the biosphere or physical environment, they will provide novel niches for microbial occupation. Examples abound, from patterns of air travel to chill compartments in supermarkets, from sexual behavior to land irrigation, from indwelling catheters to marine oil platforms, and from hide porter's disease to bovine spongiform encephalopathy. In each case, microorganisms have taken advantage of newly available conditions or substances, exploiting our designs and often thwarting our purposes.

Anyone who doubts this admittedly sweeping assertion might care to try what I have been doing: reading the journals, not from our own perspective, but from that of the microbial world. Although a mildly eccentric exercise, it provides continual insights into the constantly shifting boundaries between the micro- and macroworlds.

In a single day recently, and without consciously seeking them out, I came across three disparate examples where organisms have already gotten under our defenses and a fourth where they may have been detected just in time. They were *Mycobacterium marinum* in an inland bearded dragon, *Chlamydophila psittaci* in veterinary students

Animalcules: the Activities, Impacts, and Investigators of Microbes

and teachers, coagulase-negative staphylococci in a stem cell bank, and *Listeria monocytogenes* in ready-to-use vegetables.

The origin of the first of these is to be found in the changing tastes of pet owners. On both sides of the Atlantic, people have begun to keep in captivity the inland bearded dragon *(Pogona vitticeps)*, a so-called agamid lizard that is native to Australia's inland deserts, and the opportunist that has taken advantage of its arrival in the home is *M. marinum*, a common fish pathogen that also causes an atypical variety of tuberculosis in humans.

It was two veterinarians from the United Kingdom who discovered that *M. marinum* could proliferate in the inland bearded dragon, too. The owner of one of these creatures sought professional help because it was suffering from anorexia, lethargy, and "dullness." At that time, the owner insisted it had always been fed on a conventional diet of mealworms, crickets, and leafy greens. Only later did the owner admit to giving it dead and dying guppies from a tropical aquarium tank. Once the mycobacterium was identified in the fish, however, the cause of the dragon's malady became obvious.

"This case provides further evidence that *M. marinum* is a potential pathogen of agamid lizards, and that it appears to prefer peripheral body sites, particularly limb joints, where inflammatory effusive joint disease is seen," says a report of the case (2). "Due to the potential zoonotic nature of the disease, and the difficulty in achieving a cure in both human beings and animals, euthanasia of an animal should always be considered in cases of systemic disease."

My second invader is *C. psittaci*. Taking advantage of unsatisfactory practices in a veterinary teaching hospital in The Netherlands, it attacked both students and teachers. The occurrence came to light when both a member of staff and a veterinarian who had attended a postgraduate course at the hospital developed psittacosis. The suspected source of the organism was a flock of nine cockatiels used in a teaching session. Unfortunately, they had been brought to the hospital only for the period of the course and could not be traced thereafter or screened for *C. psittaci*.

During the session, however, instructors had also exhibited the hospital's own pigeons and Amazon parrots. The parrots were used in a later practical class, and subsequently, some of them became ill, at which point all students and staff were offered screening. When investigators tested sputum samples from patients, throat swabs from exposed students and staff, and fecal specimens from parrots and pigeons, they found that 10 out of 29 individuals were infected. Clinical features in the human patients ranged from none to sepsis with multiorgan failure requiring admission to an intensive-care unit.

Writing in the *Journal of Medical Microbiology* (3), Edou Heddema of the University of Amsterdam, together with coinvestigators elsewhere, made several practical suggestions that should prevent similar happenings and/or facilitate investigation in future. They included the maintenance of proper records regarding transactions in birds, which should also be screened for *C. psittaci* or given prophylaxis before being incorporated into a captive flock.

However, the investigators also warned of the possibility that such incursions sometimes go unnoticed. "The relatively poor control of the disease in birds and the broad spectrum of clinical syndromes in people infected with *C. psittaci* raise the question of whether this microorganism may be a much more frequent pathogen than previously recognized," they concluded. "For accurate diagnosis, we therefore recommend the detection of *C. psittaci* by PCR. . . . Sequencing of the *ompA* gene for genotyping is a helpful tool for identification of the avian source, and improves our understanding of the epidemiology of this disease in birds, humans and outbreak settings."

Next came animalcular lurkings around one of the most recent innovations in biomedical research: the establishment of stem cell banks. About 10 years ago, reports on the microbial contamination of stem cell collections began to appear. These have been followed recently by papers describing the transmission of fungi and human herpesvirus 6 to patients receiving stem cell transplants. Such findings prompted Fernando Cobo and Ángel Concha to carry out a comprehensive investigation, apparently the first of its sort, into contamination at the Stem Cell Bank of Andalucia in Spain.

The outcome, documented in *Letters in Applied Microbiology* (1), is disquieting reading. Both passive and active air sampling, together with monitoring of surfaces, revealed that the main contaminants were coagulase-negative staphylococci and other opportunistic pathogens.

The majority were normal inhabitants of the human skin, mucosa, and oropharynx, but they included organisms capable of causing bacteremia, septicemia, otitis media, and meningitis.

The two main bacteria isolated were *Staphylococcus epidermidis* and *Micrococcus luteus*, and the most critical locations in the clean rooms were a CO_2 incubator and a centrifuge. "These locations coincide with the greater work level and major presence of personnel. . . . The place of less risk of contamination was the biological safety cabinet, because of both the presence of laminar flow and the exhaustive cleanliness of this instrument."

Finally, there was a case in which both pathogens and spoilage bacteria were threatening to take advantage of the modern trade in ready-to-use vegetables. It comes from investigations by Sandra Stringer and coworkers at the Institute of Food Research in Norwich, United Kingdom. They wondered whether the heat treatment now being suggested for broccoli, lettuce, and green beans sold in this form had any drawbacks alongside the advantage of convenience. Might it damage cell membranes and inactivate plant defenses, encouraging the growth of undesirable organisms?

As reported in the *Journal of Applied Microbiology* (4), this is precisely what they found. While washing cut beans and broccoli at 52°C could reduce the growth of molds and yeasts and the development of unsightly browning, it also allowed both pathogens and spoilage organisms to proliferate more rapidly and extensively. For example, *L. monocytogenes* grew faster on vegetables washed at 52°C than on vegetables washed at 20°C.

So there you have it, representatives of four diverse groups of microorganisms probing and nibbling around human defenses in three different countries; and these came to light in a single month in the journals I peruse regularly. How many other opportunistic adventures go unreported?

References

1. **Cobo, F., and Á. Concha.** 2007. Environmental microbial contamination in a stem cell bank. *Lett. Appl. Microbiol.* **44:**379–386.

2. **Girling, S. G., and M. A. Fraser.** 2007. Systemic mycobacteriosis in an inland bearded dragon *(Pogona vitticeps). Vet. Rec.* **160:**526–528.

3. **Heddema, E. R., E. J. van Hannen, B. Duim, B. M. de Jongh, J. A. Kaan, R. van Kessel, J. T. Lumeij, C. E. Visser, and C. M. J. E. Vandenbroucke-Grauls.** 2006. An outbreak of psittacosis due to *Chlamydophila psittaci* genotype A in a veterinary teaching hospital. *J. Med. Microbiol.* **55:**1571–1575.

4. **Stringer, S. C., J. Plowman, and M. W. Peck.** 2007. The microbiological quality of hot water-washed broccoli florets and cut green beans. *J. Appl. Microbiol.* **102:**41–50.

32 Bioscience Embattled

Some scientists, not all, appear to feel that their professional lives would be more agreeable if five conditions were met. In this idealized world, journalists would always report on science accurately and responsibly, without either gloom or hype; politicians would invariably vote appropriate funds for research, including the support of untargeted basic science; regulators would always work in an objective way, resisting any temptation to respond to irrational concerns; public unease regarding developments such as genetic modification would wane in response to rational argument; and campaigners opposing those developments would go out of business.

It is easy to appreciate how this pipe dream arises. I was talking recently to a young microbiologist involved in vaccine research who was simultaneously worried about difficulties encountered in obtaining his next grant, about absurd media allegations that babies might not be strong enough to cope with a new polyvalent vaccine, about growing public disapproval of science, and about fears that his line of research might be regulated out of existence in the future.

I spoke to this man in Stockholm, Sweden, during Euroscience Open Forum 2004 (ESOF2004), billed as the first-ever pan-European general science meeting. Euroscience was established to play a role similar to that of the American Association for the Advancement of Science. It aims to influence science and technology policies in Europe, strengthen links between science and society, and provide a forum for debate on scientific issues.

Although neither Euroscience nor ESOF2004 was specifically created to address what many see as a less than ideal climate for research, many voices in Stockholm did ventilate those concerns. There were, for example, alarums over public and regulatory aspects of stem cells, transgenics, and nanotechnology and discussions based on the premise that "plant genomics are regarded with deep suspicion by many consumers and policy makers."

"The precautionary principle demands that actions be regulated on the basis of expressed concerns rather than scientific data," argued one speaker. "If precaution and safety become watchwords of our age, what is the future of scientific inquiry?" One session convener asked, "As pressure groups become increasingly savvy about publicizing scientific issues, how should the research community defend reality?" Another convener even argued that "lack of citizenship is the ordinary lot of scientists in today's Europe."

ESOF2004 embraced much more than this litany of woe. There were celebrations of achievements in fields from particle physics to molecular evolution, from psychology in business to biology in silico. Excellent, wide-ranging debates embraced global warming, sustainable development, the demographic challenge, and much else, and an outreach program brought science to all ages through exhibitions, expeditions, excursions, a science cafe, and the Klara Soup Theatre ("Physics and love meet in this story about a remarkably obstinate scientist").

Yet angst was omnipresent, too. "Why is Europe giving up its strong position in pharmaceutical R&D?" asked one speaker. Another inquired, "Why should the public influence the scientific agenda in neurobiology?" while another asked, "Can chemistry survive in Europe?"

I am conscious of the real problems that have given rise to pessimistic thoughts of this sort. It is certainly not difficult to sympathize with the young microbiologist I met in Stockholm. Though some of his fears (such as public disenchantment with science) were exaggerated, he was understandably dismayed over perceived threats to branches of science specifically concerned with improving the quality of life. "This antiscience stuff seems to have got out of hand," he said.

He had a point, yet it is also worth examining the likely consequences if all of the tensions and conflicts now surrounding science (especially in Europe) were to be totally resolved. Would science, and society at large, be true beneficiaries? Or is it possible that, despite all of our present difficulties, an adversary relationship is healthier than consensual tranquility?

Take the oft-lamented relations between science and the media. Do researchers really want journalists to behave purely as scribes, writing down what they are told and seeing that it is published accurately and colorlessly? Scientists who fancy this idea should consider whether they wish reporters to show the same subservience when writing about the work of a rival, especially a heterodox one.

I can recall many occasions when pharmaceutical companies, in particular, have been infuriated by uncritical media coverage of their competitors' products. I remember, too, the differing ways in which London-based journalists, many years ago, dealt with Linus Pauling's claims that daily megadoses of vitamin C could combat both the common cold and cancer. Some reported these assertions, bolstered by the enormous distinction of a double Nobel laureate, without caveats or context. Others very properly coupled them with quotes from skeptical researchers and reminders that the claims had not been validated by independent studies.

Even the color and impact of a newspaper story often rest upon a challenging interplay between reporter and scientist. This is what turns a dull and unduly precise comment into an enticing, readable one. Compare "It has been reported that the drug is effective in reducing the clinical severity of cold symptoms in 72.3% of cases" with "The drug relieves cold symptoms in three-quarters of people."

The *New England Journal of Medicine* once complained about "gene cloning by press conference," the announcement and reporting of claims, in the early days of recombinant DNA, regardless of their real significance. It was, however, the companies courting the media in this way, rather than the journalists, who were the journal's principal target. Another offense, even today, is the reporting of bioremediation claims based solely on laboratory scale experiments, when journalists fail to question claimants about the feasibility or likely economics of a practical process.

As for the unreasonable influence on science of lobbyists and the public, it is as well to remember that they are not always wrong (nor scientists always right). Think of Old Lyme, CT, when worried parents in the 1970s were crucial in calling attention to an "epidemic of arthritis" and insisting that it be investigated. Dismissed by experts initially as nonsense, the agitators' claims actually led to the recognition and management of what we now know as Lyme disease (see chapter 30).

On occasions of this sort, campaigners and the public are rarely, if ever, clamoring for hocus-pocus. They are asking for real, objective scientific investigation. Think of the AIDS marches when that disease first emerged. Even the opponents of novel vaccines or genetically modified foods invariably insist that "more research is needed." They do not want aromatherapy or medieval agriculture.

Finally, as regards assertions that research is now overregulated, a little history is helpful in providing perspective on today's preoccupations. Does anyone now believe that the post-Asilomar 1970s guidelines and regulations on recombinant DNA experiments, then fiercely opposed, were anything but prudent and necessary? What of the new wave of genetic engineers at that time, throwing their *Escherichia coli* cultures down the sink without regard for the sort of precautions long practiced by medical microbiologists?

This chapter has, of course, been a somewhat indulgent exercise in special pleading. I have ignored many causes of genuine grievance among scientists, from journalistic irresponsibility to mindless meddling by politicians. However, it is probably more realistic to see these things as inevitable accompaniments of a social setting for science that is basically open and healthy than to crave a utopia that is never going to exist anyway, and that setting is characterized by competition, conflict, and tension.

Perhaps organizations such as Euroscience, while rightly highlighting the immense, unique practical and cultural value of science, could recognize this wider reality more explicitly in their deliberations in future. It was certainly unwise, on the occasion of ESOF2004, for Euroscience's president to list as the first of its objectives the securing of better pension rights for scientists.

33 "Playing God"

"Milk which has been pasteurized at 165°C is more liable to induce scurvy than either boiled milk, or milk which has been pasteurized at lower temperatures, such as 140–145°C for 30 minutes." So wrote the distinguished Johns Hopkins biochemist Elmer McCollum in his book *The Newer Knowledge of Nutrition: the Use of Food for the Preservation of Vitality and Health,* published in 1918 (2). His remarks came from a wide range of science, erroneous science, pseudoscience, prejudice, and vested interests, which delayed for decades the wide-scale introduction of pasteurization in both the United States and Europe. Resistance to the practice endured long after the emergence of compelling evidence of both its safety and its power to prevent lethal milk-borne infections.

I doubt whether 21st-century opponents of genetically modified (GM) crops, or indeed GM organisms in general, or xenotransplantation, or therapeutic cloning, or stem cell research pay much attention to history. They should do so. There are instructive parallels between many past campaigns against biomedical innovations and those of today.

Motifs common to both include allegations of scientists "interfering with nature" or "playing God"; claims that only commerce, not consumers, will benefit; the invention of bogus hazards; and wild exaggerations of potentially real risks, which we are said to be powerless to contain. Opponents also insist that applications must be banned until long-term studies have been completed.

Of course, such studies will themselves also be resisted. In Europe, some activists have attacked field experiments specifically designed to evaluate the environmental impact of GM crops. These even include plants developed for their potential environmental benefits—poplar trees, for example, containing less lignin than usual, which means lower temperatures and chlorine levels in paper making.

Most campaigners will dismiss this analysis as a caricature. Nevertheless, looking back over my lifetime and beyond, I see many more occasions when pioneers have been proved right than when their opponents have prevailed. Two come from the field of tissue transplantation, with its strong links to the science of immunology.

Just over half a century ago, corneal grafting had become a practical possibility. However, its use in Britain to save patients' eyesight was forbidden by Anatomy Acts placed on the Statute Book in 1832 and 1871. There were doubters and critics, too, who saw the potential new operation as dangerous and unnatural. Fortunately, reason prevailed, thanks to potent lobbying by a distinguished trio: an ophthalmic surgeon, an influential journalist, and a sympathetic member of parliament. The law was amended in 1952.

Fifteen years later, Christiaan Barnard performed the world's first heart transplant in Cape Town, South Africa, provoking a storm of criticism throughout the world. According to the critics, "heroic" surgery of this sort was contrary to nature or the Bible. It was irresponsible in light of the imperfect nature of immunosuppression at that time. It represented a gross distortion of medical priorities, draining away vast resources from more needy areas of health care, and in any case, the operation would be used to rescue people whose cardiovascular problems stemmed from smoking, obesity, or eating junk food.

Today, both corneal and cardiac transplantations are so well established that it is difficult to believe the strength of opposition surrounding their inceptions. Blindness is being prevented at all ages, and children with new hearts (even hearts and lungs) now compete in the Transplant

Animalcules: the Activities, Impacts, and Investigators of Microbes

Olympics. Their cardiac problems were not caused by unhealthy living, but in many instances by congenital conditions.

These are the realities which present-day campaigners against various applications of biotechnology may care to contemplate. One of several lessons they might draw is that genuine technical problems should not be exaggerated or used to support objections that are really dogmatic in origin.

History also shows that technical difficulties are solvable, whereas those rooted in fundamentalism are by definition insurmountable. All scientific advances bring genuine technical challenges in their wake, but the difficulties over immunosuppression that bedeviled organ grafting in the 1960s were largely conquered. Likewise, those posed by pig endogenous retroviruses for xenotransplantation today will be surmounted in one way or another.

Joseph Hotchkiss of Cornell University has highlighted the controversies that invariably surround novel technology with reference to the introduction of milk pasteurization a century ago. Far from being wholesome, milk was then a source of tuberculosis, typhoid fever, scarlet fever, diphtheria, Malta fever, and many other infections, yet the benefits of Louis Pasteur's eponymous process in combating the pathogens responsible for these diseases were abundantly clear. When compulsory pasteurization was introduced in New York City between 1899 and 1910, the infant mortality rate plummeted from 12 to 3.8 per 1,000.

Despite evidence of this sort, the opponents were remarkably diverse, in contrast to the professional lobbyists of today. "For decades, strong and adamant opposition succeeded in stalling moves to make pasteurization mandatory in many parts of Europe and in North America," Hotchkiss wrote (1). "The opposition came from almost all quarters, including the medical community, the dairy industry, dairy technologists, and the milk-consuming public."

Two of their arguments were that pasteurization would mask low-quality milk and that it was unnecessary if the milk was properly handled so that it was no longer a source of infection. There are strong echoes here of contemporary objections to food irradiation, which could prevent *Salmonella*, *Campylobacter*, and other infections on a scale similar to that already achieved by pasteurization. The parallel is compelling: an allegation that appears reasonable at first glance but that a moment's examination shows to be spurious.

Joseph Hotchkiss made a damning comparison between the failures of innovators and the scientific establishment in the early decades of pasteurization and those grappling with hostility toward genetic modification in the early 21st century. "Anticipation and planning should accompany technological development, and not be a reactive response," he wrote. "When controversy is not anticipated and planned for, technologists and scientists are forced into the position of reacting to the debate as framed by others, rather than being framers of the debate."

A specific example of bioscientists' failure to discern the form of future public debate is afforded by the use of antibiotic resistance genes as markers in the production of transgenic plants. The researchers who first applied this technique saw it simply as a convenient laboratory tool. They did not anticipate that the possibility of such resistance being transferred to pathogens would emerge later as a major public safety concern (upon which even regulatory committees have taken differing positions).

"The controversy surrounding pasteurization also points to the importance of the media. . . . It is important to educate the media early in the development stage and not to delay until implementation," wrote Hotchkiss. Here, I enter pleas of both support and dissent. The importance of media relations in areas of emerging technology can hardly be overstated: this is where many of the mistakes regarding GM food have been made.

However, well-meaning experts need to recognize that journalists see themselves as neither recipients nor purveyors of "education." They are as uneasy with the notion of being "educated" by scientists as they are with that of "educating" their readers. In both cases, a little subtlety is required.

Whether we call it education or simply the dissemination of information (as I much prefer), a cardinal error in the GM food debate was the failure of proponents to put the case for GM crops on the record sensitively and at the right time (which means early in the 1990s). In consequence, the subject was allowed to become a wholly negative item on the agenda for public discourse. At least in Europe, claims about environmental and nutritional benefits were interpreted as defensive attempts at rationalization and were largely ignored.

No doubt we shall eventually reap those rewards, just as we benefit today from milk pasteurization, but it is sad that we have learned so little from history in preparing the way for new technology.

References

1. **Hotchkiss, J. H.** 2001. Lambasting Louis: lessons from pasteurization, p. 51–68. *In* A. Eaglesham, S. G. Pueppke, and R. W. F. Hardy (ed.), *Genetically Modified Food and the Consumer*. National Agricultural Biotechnology Council, Ithaca, NY.

2. **McCollum, E.** 1918. *The Newer Knowledge of Nutrition: the Use of Food for the Preservation of Vitality and Health*. Macmillan, New York, NY.

34 Microbes in the Media

The coronavirus responsible for severe acute respiratory syndrome (SARS) may yet have more lessons in store for us. Thus, it would be foolish to dismiss as totally baseless the apocalyptic headlines that swept the world for several months after the disease first appeared in China in 2002–2003. Nevertheless, I sympathize with Martin Kelly's suggestion (2) that the high profile given to SARS was disproportionate when considered alongside other, similarly dangerous conditions.

Kelly, from Altnagelvin Area Hospital in Derry, Northern Ireland, highlighted in particular severe community-acquired pneumonia (CAP). The contrast between the two infections is certainly instructive. SARS did not cause the millions of deaths originally forecast throughout the planet. It killed a total of 774 people in 29 countries, and its damaging economic effects (on tourism, for example) stemmed as much from hysterical overreaction as from the activity of the virus itself. By contrast, Kelly said, "countless numbers of cases of CAP often go unrecognized and unpublicised. These too may be severe, life-threatening or even fatal."

Since any newly recognized syndrome or unfamiliar pathogen just might represent the advent of a fearsome pandemic, it is not easy to get these things right. Still, journalists, and sometimes health authorities,

Animalcules: the Activities, Impacts, and Investigators of Microbes

too, could do more to encourage a climate of realism, especially at the outset, when an epidemic is suspected.

Key questions are whether the disease and/or the organism is genuinely new. In May 1994, for well over a week, Britain was seemingly attacked by a "flesh-eating virus" that was "rampaging throughout the country." It was "resistant to all known antibiotics" and caused a condition "hitherto unknown to medical science." All of these newspaper assertions were incorrect.

The reality was an antibiotic-sensitive hemolytic streptococcus. The infection was the equally familiar necrotizing fasciitis. There was no epidemic or significant cluster of cases, simply the normal, tiny incidence among patients at exceptional risk. However, these facts were not publicized for many days. One reason was the reluctance of bodies such as the Medical Research Council to provide journalists with authoritative information. Horrified by media sensationalism, the Medical Research Council recoiled into silence.

What journalists require on these occasions (in order to inform their readers and listeners) are answers to simple, practical questions. What is already known about the organism, especially its virulence and transmissibility? How long will it take to find out things that are not yet known (for example, how it spreads and what treatments are available)?

Often, amid the panic, these matters are overlooked. In some countries, for example, it was many weeks after the first scary headlines announcing the eruption of SARS before the issue of possible person-to-person transmission was broached, yet this was crucial in assessing the potential scale and seriousness of any outbreak.

One frequent misrepresentation of the threats posed by newly recognized infections and novel manifestations of familiar ones comes from use of the expression "superbug." I have even heard microbiologists brandishing this term when they mean either an exceptionally virulent organism or a multiply resistant organism, or even a highly transmissible one. Not all journalists are aware that these are independent traits, which is why superbugs often appear in the media as apparently possessing all three.

Living in the United Kingdom, I was reminded several times during the SARS affair of our national hysteria over the typhoid fever outbreak that occurred in Aberdeen, Scotland, in 1964. Everyone of the appropriate age recalls that incident vividly. They do so not because it was exceptionally serious, but largely because of manic coverage in the press, unwittingly encouraged by Aberdeen's Medical Officer of Health (MOH).

Writing in *Epidemic Diseases in Aberdeen and the History of the City Hospital* (3), Ian Porter and Michael Williams pointed out some of the mistakes that were made and the repercussions.

At the request of the MOH, names and addresses of patients were published. . . . City food shops and restaurants reported a slump in trade and several food production plants were closed. Many sporting fixtures and outings were cancelled. All city schools were closed as was the College of Domestic Science. Dance halls in the town were also shut and the MOH criticised 'selfish citizens' who were still attending dances out of town.

The city streets were all washed down with disinfectant. Attendances at cinemas slumped and hoteliers reported many cancellations of bookings by visitors. There were reports of Aberdeen citizens being denied entry or even turned away from hotels elsewhere in the country.

Tourists on a ship from Finland which called at Aberdeen harbour stayed aboard. Employees of a Perth firm who were working in Aberdeen were asked not to return home at weekends. Some travel agencies would only accept bookings from Aberdonians if they were immunised against typhoid. . . . Speyside Council, on the advice of their MOH, banned Aberdeen citizens from visiting their area.

The university issued a circular to all outside examiners recommending that they wear cotton gloves when marking scripts from the city. An outside library also requested that a book which had been lent to Aberdeen University library be fumigated before being returned, and requested a certificate to certify that having been done.

Many of these actions were unnecessary, ineffective, and/or counterproductive. Publication of names and addresses of patients, for example, led to demands for tests by many people who were in no sense at-risk contacts, and the initiative by travel agencies created a situation in which recently immunized individuals who coincidentally happened to become ill could be misdiagnosed as typhoid fever cases because they would show rising antibody titers.

Health authorities today might learn from the Aberdeen experience—not from its more comical features, perhaps, but from the needless use of measures that were based on an erroneous understanding of the communicability of the disease. The MOH, who had attacked the media for its sensationalism, was himself heavily criticized by the subsequent official inquiry. The report also concluded that the city's publicity campaign had been of little value in containing the outbreak.

Still, some good did come out of Aberdeen: new knowledge about *Salmonella enterica* serovar Typhi plus insight into another epidemic that occurred many years previously. In Aberdeen, suspicions regarding the source of the organism centered on a can of corned beef from a local store. It appeared that *S. enterica* serovar Typhi had entered the can through a faulty seal during cooling with sewage-contaminated water after the can had been autoclaved at the manufacturing plant in Argentina.

However, there was considerable disbelief that the typhoid bacillus could have survived in such a can. Even if it had entered in this unlikely way, critics argued, it could not compete with bacteria, such as *Escherichia coli*, that would have accompanied it in greater numbers. The bacilli would also betray their presence by "blowing" the cans and/or discoloring the meat.

E. S. Anderson and Betty Hobbs of what was then called the Enteric Reference Laboratory in London demolished this theory. First, they found that the Aberdeen strain of *S. enterica* serovar Typhi grew more prolifically than *E. coli* in corned beef. When tested in cans of corned beef of the same batch as that suspected to have caused the Aberdeen outbreak, *S. enterica* serovar Typhi multiplied and spread.

It could be isolated from both ends of the meat after 3 years of storage and was still widely distributed even after 8 years. Moreover, cans infected with *S. enterica* serovar Typhi, together with *E. coli* and *Enterobacter cloacae*, were not blown, and the meat was not visibly spoiled. The typhoid bacillus thrived better than the other two organisms.

This explanation, combined with some later detective work by Anderson, almost certainly also accounts for a typhoid fever outbreak in Oswestry, United Kingdom, in 1948 (1). That outbreak, which affected almost exactly the same number of patients as in Aberdeen, attracted virtually no coverage in the national press. Britain was far too preoccupied with the aftermath of war.

References

1. **Anderson, E. S., and B. C. Hobbs.** 1973. Studies of the strain of *Salmonella typhi* responsible for the Aberdeen typhoid outbreak. *Isr. J. Med. Sci.* **9:**162–174.

2. **Kelly, M. G., R. A. Sharkey, K. W. Moles, and J. G. Daly. 2004.** Severe community-acquired pneumococcal pneumonia (CAP)—a potentially fatal illness. *J. Med. Microbiol.* **53:**83–84.

3. **Porter, I., and M. Williams.** 2001. *Epidemic Diseases in Aberdeen and the History of the City Hospital.* Aberdeen History of Medicine Publications, Aberdeen, United Kingdom.

35 A Little Learning…

"Scientific endeavour is almost epitomised by the white coated scientist. Yet those same white coats bring their unexpected problems for laboratories dealing with genetically engineered bacteria as they provide a neat escape route for the bugs." This passage from the Greenpeace report *Genetic Engineering: Too Good To Go Wrong?* (3) is truly (and no doubt intentionally) alarming. Scientists working with genetically modified organisms are, it seems, capable of spreading them into the outside world. Greenpeace went on to tell us that genetically modified microbes splashed onto lab coats were hitherto thought to dry out and die. Now we knew that they remained viable.

There was worse to come. The report argued that the first step in laundering, a soak at 35°C, is "just right for releasing the bacteria, which are then flushed into the sewerage system." This showed that "apparently great precautions in the use of genetically engineered organisms can be totally undermined by the 'oops, we didn't think of that!' factor."

The reality behind this scary scenario was a paper published by Hayo Canter Cremers and Herman Groot of the National Institute for Public Health and the Environment in The Netherlands (1). They recovered *Escherichia coli* K-12 from laboratory coats and hypothesized that it might survive washing and thus enter the sewers.

However, it was scarcely surprising that *E. coli* K-12 (which was, of course, widely used as an innocuous research tool long before the advent of recombinant DNA) could sometimes be found on laboratory clothing. More importantly, white coats could hardly be expected to serve as barriers to prevent genuinely hazardous organisms from being disseminated and causing public danger.

In suggesting that this is their purpose, and that these safeguards are regularly breached, Greenpeace was being mildly ridiculous. No one in their right senses has ever believed that pathogens can be prevented from escaping from laboratories by the outer garments worn by scientists and technicians, nor even by scrupulous laundering. What fulfils this purpose instead is containment and other safety procedures, backed by laws and guidelines. These are not "totally undermined" by the predictable results of a few experiments with lab coats.

Why do I waste column inches in discussing Greenpeace's alarm? Simply to show that, if newspaper editors and their readers stopped to think for one moment, they should be able to recognize an argument of this sort for what it really is. The same applies to many other microbiological issues that surface regularly in the media.

However, spotting such animalcular fallacies would require greater microbial literacy than is yet commonplace. Microbial literacy would mean knowing that our clothing is heaving with microorganisms from our bodies and environment. It would mean understanding that mouth washes cannot render the oral cavity "germ free" for more than a few minutes at a time, that there are many more beneficial microorganisms in the world than harmful ones, and that the bacteria we carry, rather than the body itself, can become resistant to antibiotics.

It would mean understanding that simple principles of food hygiene in the kitchen are far more important than deploying household antiseptics to kill "all known germs"; that we are not all riddled with, and debilitated by, *Candida;* and that the late Princess Diana did not need to shake hands with an AIDS patient to establish that this was not a route of infection. The list of similar misconceptions is endless and, in an era of biotechnology, growing.

However, comparatively little has been done to improve matters. True, there are some splendid exceptions, including Douglas Zook's Microcosmos at Boston University and the American Society for Microbiology's Microbial Literacy Collaborative. Nevertheless, many initiatives to promote the public understanding of science have failed to recognize the exceptional importance of microbiology as a discipline underlying so many social issues today.

Indeed, measures to promote and measure scientific literacy can themselves be flawed, reflecting an undue haste to categorize members of the public as either well informed or ignorant. I do not exclude my own comments above from the cautionary criticism that shaky assumptions may underlie efforts to assess and improve public understanding of science.

Consider, for example, the European Commission report *The Europeans and Modern Biotechnology,* published in 1998 (2), which presented the conclusions of a Eurobarometer opinion poll conducted in the 15 member states of the European Union. This was a valuable document, revealing a comparatively optimistic view of the developments people expect to stem from biotechnology, though showing considerable variations between one country and another. One of its most telling findings was that "optimism" was closely linked with objective knowledge of the subject, as was "pessimism," to a lesser degree. The percentage of "don't knows" declined steeply with increasing knowledge.

Clearly, the methodology used to find out what participants understood about biotechnology and related disciplines was crucially important. Here, I venture some anxieties. There were 10 statements, and respondents were asked to determine whether each of them was true or false. They could also answer "don't know," but no other reply was permitted.

One statement was as follows: "Viruses can be contaminated by bacteria." I have tested this on eight qualified microbiologists, none of whom reacted appropriately according to the Eurobarometer. Two refused to answer at all, insisting that the question was as meaningless as "Can apples be contaminated by pears?" One reluctantly opted for "don't know."

The remaining five, with varying degrees of unease, selected true on the grounds that tissue cultures of viruses may actually be contaminated with bacteria. That is why antibiotics, such as benzyl penicillin, gentamicin, and tetracycline, are incorporated in cultures to keep bacteria at bay. However, all of this would be irrelevant to the Eurobarometer subjects, who were judged to be knowledgeable only if they answered "false."

Another statement was "Cloning living things produces exactly identical offspring." Again, a quick check with half a dozen biological colleagues revealed that they all failed (according to the Eurobarometer) by selecting "true."

There is, of course, room for uncertainty in the expression "living things," and in the case of microorganisms, the statement could be considered to be correct. Nevertheless, public discussion and media coverage of cloning invariably focus on animals, including humans. This is the area that has seen the overwhelming weight of debate over the years since the birth of the sheep Dolly at the Roslin Institute in Scotland in 1997 and the ensuing worldwide calls for bans on human cloning. In this case, the correct answer was certainly "false."

Dolly carried her father's nuclear DNA but her mother's mitochondrial DNA. This, combined with subsequent nurture and other factors, means that such clones are not exactly identical. Following the appearance of Dolly, researchers and bioethicists repeatedly emphasized that the specter of identical Hitlers is just that, a phantasm with no reality.

What, then, are we to make of the fact that 48% of Eurobarometer respondents were judged to be incorrect in believing that viruses can be contaminated by bacteria? What of the 46% given full marks for agreeing that cloning living things produces exactly identical offspring? How to assess these results alongside the 68% who agreed on a proposition on which both the Eurobarometer analysts and all microbiologists would also concur, that "Yeast for brewing beer contains living organisms"? Overall, what conclusions can we reasonably draw from such soundings regarding peoples' biological knowledge or ignorance?

Pollsters should perhaps consider whether the hallmark of a (microbiologically) literate person is their refusal to provide black and white answers to questions of this sort. This may well make it more difficult to design easy-to-complete questionnaires to evaluate scientific literacy in the future, yet reality really is more complicated than "true or false" would suggest—unless you happen to be Greenpeace.

References

1. **Canter Cremers, H., and H. Groot.** 1991. *Survival of* E. coli *K12 on Laboratory Coats Made of 100% Cotton.* RIVM report no. 719102009. National Institute for Public Health and the Environment, Bilthoven, The Netherlands.

2. **European Commission.** 1998. *The Europeans and Modern Biotechnology.* European Commission, Brussels, Belgium.

3. **Greenpeace.** 1997. *Genetic Engineering: Too Good To Go Wrong?* Greenpeace, London, United Kingdom.

36 Spotlight
   on Acetaldehyde

Everyone knows that excessive drinking causes esophageal cancer because alcohol, especially in neat spirits, acts as a carcinogen on the cells lining the esophagus. Smoking is also implicated. It seems that little more needs to be said.

Mikko Salaspuro of the University of Helsinki in Finland disagrees. He believes that microorganisms play a crucial role, too, and his somewhat heretical viewpoint is backed by abundant evidence (mostly published in journals devoted to alcohol and cancer rather than microbiology). Salaspuro argues that bacteria in the saliva and elsewhere in the digestive tract convert alcohol to acetaldehyde, which then acts as the real carcinogen. He believes that certain individuals may harbor organisms that are particularly active in forming acetaldehyde. Does this indicate a potential avenue for prevention or treatment?

The idea is appealing, not least because, despite epidemiological evidence linking heavy drinking with certain cancers, oncologists do not regard alcohol as a true carcinogen. Acetaldehyde is so regarded, because it induces chromosomal aberrations, micronuclei, and/or sister chromatid exchanges in cultured mammalian cells and gene mutations at the *hprt* locus in lymphocytes in vivo.

About 15 years ago Salaspuro, a gastroenterologist and substance abuse expert, first became interested in the possible activities of drinkers' salivary animalcules in producing carcinogenic acetaldehyde. Because ethanol is rapidly and evenly distributed to the water phase of all organs in the body, his interest extended to the microbial populations of the colon, too.

Salaspuro and his colleagues discovered that normal human colonic contents, especially their aerobic bacteria, generated significant quantities of acetaldehyde when incubated aerobically or microaerobically with ethanol in vitro. Writing in the *Scandinavian Journal of Gastroenterology* (4), they suggested that this bacterial adaptation might be an essential feature of what they called the bacteriocolonic pathway to form acetaldehyde from exogenous (or endogenous) alcohol.

Working with piglets, they found that high levels of acetaldehyde were formed in this way in the colon. Moreover, experiments on rats established that ciprofloxacin, which suppressed the aerobic and facultatively anaerobic flora, prevented acetaldehyde from accumulating in the colon. Metronidazole, on the other hand, caused the same populations to proliferate, with a concomitant steep rise in acetaldehyde.

Another type of evidence comes from studies on "Oriental flushers." These are individuals with genetically deficient aldehyde dehydrogenase (ALDH2) whose faces redden after they consume alcohol; the partially inactive enzyme cannot deal quickly enough with acetaldehyde produced by alcohol dehydrogenase (ADH). Several recent epidemiological surveys have shown that those who flush but nevertheless abuse alcohol have significantly heightened risks of cancers of the esophagus, colon, and other parts of the intestinal tract.

Writing in *Alcoholism: Clinical and Experimental Research* (5), Salaspuro and coworkers reported that, following a moderate dose of alcohol, Asians with the mutant ALDH2 developed salivary acetaldehyde levels two to three times higher than those in Asians with normal ALDH2. They argued that, combined with epidemiological data, this finding provided strong evidence for the oral carcinogenicity of acetaldehyde produced in the saliva by microbes in the mouth or salivary glands.

Other papers by the Finnish group have demonstrated that, in addition to tissue ADH, gastrointestinal tract bacteria and the microflora of the mouth can oxidize ethanol with their ADH. As a consequence, very high concentrations of acetaldehyde occur in the saliva and in the contents of the large intestine during the oxidation of alcohol.

As a means of preventing the adverse consequences of this accumulation, Salaspuro and colleagues have experimented with the nonessential amino acid L-cysteine. This reacts covalently with acetaldehyde (to form 2-methylthiazolidine-4-carboxylic acid). In principle, therefore, it could be used in slow-release tablets to forestall carcinogenicity.

One report from the Helsinki team in the *International Journal of Cancer* (3) established that such tablets, given to human volunteers, are indeed highly effective in removing acetaldehyde from the saliva. Cysteine also binds acetaldehyde in the colon, while lactulose can be used to diminish its microbial production in that area.

The microbial theory could also help to explain why people who do not take care of their teeth have an increased risk of cancer of the oral cavity. Although this association is well established, there is uncertainty about the underlying reason(s). Some evidence points to a synergistic effect between heavy drinking and poor dental hygiene, just as there is between alcohol abuse and smoking.

Could the microbial generation of acetaldehyde be the missing link? The Finnish researchers sought to answer this question by studying over 100 volunteers, who differed not only in their drinking and smoking habits, but also in the quality of their dental hygiene. Saliva samples were collected and assayed for the amounts of acetaldehyde formed when they were incubated with ethanol. The results, reported in *Oral Oncology* (1), showed that poor dental hygiene was associated with a twofold increase in acetaldehyde production.

A further strand of evidence comes from the comparison of saliva from different individuals who varied in their capacities to produce acetaldehyde. "*Streptococcus salivarius*, alpha-hemolysing *Streptococci, Corynebacterium* spp and *Stomatococcus* spp are the candidate bacteria potentially causing this effect," wrote Salaspuro and his collaborators in *Oral Oncology* (1).

"The observation of increased yeast colonisation in patients with poor dental hygiene might be of special interest since several *Candida* strains have been shown to occur more frequently and in higher densities in risk groups of oral cancer. Moreover, these species had a significantly higher acetaldehyde production capacity."

Another organism identified, this time by a Japanese group, as a major contributor to acetaldehyde formation in the upper digestive tract is *Neisseria*. Manabu Muto and coworkers in Tokyo and Kashiwa found that an organism of this genus produced more than 100 times greater quantities of the carcinogen when incubated in medium containing ethanol than other organisms isolated from the mouth. "Although *Neisseria* present in the normal oral microflora is generally non-pathogenic, these findings suggest that this microbe can be a regional source of carcinogenic acetaldehyde and thus potentially play an important role in alcohol-related carcinogenesis," they wrote in the *International Journal of Cancer* (2).

Speaking in Brussels in 2003 at a conference held by the alcohol industry's International Medical Advisory Group, Mikko Salaspuro said that poor dental status, individual differences in gut microflora, smoking habits, genetic factors, and heavy drinking may all modify the local formation of carcinogenic acetaldehyde in the digestive tract. The microbial production of acetaldehyde from alcohol was the major source of salivary, intragastric, and intracolonic acetaldehyde, though the salivary glands and mucous cells may sometimes play a substantial role.

Are we about to see a new, microbiological approach to the pathogenesis and prevention of alcohol-associated digestive tract cancers?

References

1. **Homann, N., J. Tillonen, H. Rintamäki, M. Salaspuro, C. Lindqvist, and J. H. Meurman.** 2001. Poor dental status increases acetaldehyde production from ethanol in saliva: a possible link to increased oral cancer risk among heavy drinkers. *Oral Oncol.* **37**:153–158.

2. **Muto, M., Y. Hitomi, A. Ohtsu, H. Shimada, Y. Kashiwase, H. Sasaki, S. Yoshida, and H. Esumi.** 2000. Acetaldehyde production by non-pathogenic *Neisseria* in human oral microflora: implications for carcinogenesis in upper aerodigestive tract. *Int. J. Cancer* **88**:342–350.

3. **Salaspuro, V., J. Hietala, P. Kaihovaara, L. Pihlajarinne, M. Marvola, and M. Salaspuro.** 2002. Removal of acetaldehyde from saliva by a slow-release buccal tablet of L-cysteine. *Int. J. Cancer* **97**:361–364.

4. **Salaspuro, V., S. Nyfors, R. Heine, A. Siitonen, M. Salaspuro, and H. Jousimies-Somer.** 1999. Ethanol oxidation and acetaldehyde production in vitro by human intestinal strains of *Escherichia coli* under aerobic, microaerobic, and anaerobic conditions. *Scand. J. Gastroenterol.* **34:**967–973.

5. **Väkeväinen, S., J. Tillonen, D. P. Agarwal, N. Srivastava, and M. Salaspuro.** 2000. High salivary acetaldehyde after a moderate dose of alcohol in ALDH2-deficient subjects: strong evidence for the local carcinogenic action of acetaldehyde. *Alcohol Clin. Exp. Res.* **24:**873–877.

37 Measles, Polio, and Conscience

In 1913, building on ideas going back to the 17th century, the Austrian mystic Rudolf Steiner promulgated his own brand of "anthroposophy," whose tenets ranged from the rejection of chemicals in agriculture to the alleged therapeutic benefits of music and colored lights. His worldview has endured, and today there are communities throughout the world based on Steiner's pantheistic speculations.

Why should this be of concern to readers today? Because of the impact of these strange notions on health. Like many other belief systems, anthroposophy has had mixed consequences. Its schools and clinics, not only in Europe, but also in North and South America, have acquired a reputation for helping mentally handicapped children and others with special needs. On the other hand, Steinerian communities tend to reject vaccination against infectious diseases.

As expounded in *Physiology and Therapeutics* (3), Steiner believed that febrile illnesses, such as measles and scarlet fever, were related to a child's spiritual development. Following this line of thinking, present-day communities founded upon his approach do not object to immunization simply as part of a wider antipathy to conventional medicine. Adherents also believe that the use of vaccines (particularly the measles vaccine) deprives infants of the opportunity to benefit from the experience of having the diseases.

Against this background, it is no surprise (though still a shock) to find that the first nationwide outbreak of measles in Great Britain since the implementation of mass vaccination over 3 decades ago occurred within nonimmune anthroposophic communities. It came to light shortly after a 5-year-old boy from a Camphill community in Yorkshire, in northern England, developed measles following a visit to a similar community in north London. Although measles had not been confirmed there by laboratory tests, about 30 of the children showed the typical rash and fever of the disease.

More cases soon began to appear in the Yorkshire community, which meanwhile was visited by an unimmunized family from another anthroposophic group in Gloucestershire. The children in that family, too, developed symptoms of measles after returning home, triggering off more new infections. Eventually, nearly 300 individuals were affected.

Of 46 salivary samples which investigators were able to obtain from the Yorkshire cluster, plus 99 samples from Gloucestershire, 117 were positive for measles. There were also a further 26 linked cases in other Steinerian communities. Overall, only two of the victims had been vaccinated against the disease.

Fortunately, this outbreak did not spread beyond the anthroposophic communities where it began. As the investigators pointed out (2), this was undoubtedly a consequence of the high level of measles immunization in Britain over many years. Following the initial introduction of a single-antigen measles vaccine in 1968, a combined measles-mumps-rubella vaccine has been widely used. At the time when the outbreak in the Steinerian communities erupted, coverage in the general population among children aged 2 years had been over 90% for 6 years.

In unimmunized societies, however, the disease still typically occurs in epidemics every 2 to 3 years. It can cause devastating encephalitis and other complications and, indeed, remains a killer in many parts of the world. Every year, measles virus kills about a million children, mainly in developing countries.

Principled objections to immunization are, of course, not unique to Britain or any other country. Daniel Feikin and colleagues have pointed out (1) that 48 U.S. states permit "religious" exemptions and 15 states permit "philosophical" exemptions from mandatory vaccination. That is one solution to the dilemma faced by authorities who wish to enforce immunization laws and yet respect the concerns of citizens opposed to their enforcement.

However, the practical consequences of parents' declining protection for their children are all too apparent. Feikin et al. cite evidence that the risk of measles infection from 1985 to 1992 in the United States was on average 35 times higher in children with personal exemptions than in vaccinated children. Likewise, countries where there are more active antivaccine movements have higher rates of pertussis than those where the majority of youngsters are immunized.

The other vaccination scandal in Europe, particularly in The Netherlands, concerns poliomyelitis. Between September 1992 and February 1993, 71 individuals in The Netherlands contracted the disease. Two died, and 59 were paralyzed. All but one belonged to an extreme Protestant sect that rejects immunization.

The World Health Organization (WHO), which in 1988 had proclaimed its hope of ridding the world of polio by 2000, was so appalled by the incident that it issued a press release pointing out the dangers posed by small pockets of unprotected individuals. While praising the Dutch authorities for containing the outbreak, the WHO warned that such incidents were obstacles on the road to global eradication.

The 1993 outbreak was simply part of a continuing pattern in The Netherlands since the inception of immunization in 1957. Although not compulsory, the vaccine was soon widely accepted, and coverage rose to about 97%. Nevertheless, the 1960s and 1970s saw several outbreaks. A particularly large one affected 110 patients in 1978. All of these epidemics were confined to people living in sectarian communities who belonged to orthodox reformed churches whose members declined protection for themselves and their children.

An intriguing aspect of the Dutch situation concerns that country's choice of polio vaccine. In the United States and many other countries, the Sabin (live attenuated) version had gradually but completely replaced Salk (inactivated killed) vaccine by the late 1960s. In contrast, The Netherlands has continued to use the Salk vaccine. This decision was based on the tiny but nevertheless real possibility that the attenuated virus can revert to virulence.

One of the merits of attenuated virus is that vaccinated children shed it in their feces and inevitably pass it on to others in nurseries and elsewhere. There is, in consequence, a good chance that some infants who have not been formally immunized will acquire the organism passively and thereby become immune. This does not, of course, happen when inactivated killed virus is administered instead.

It is at least possible, therefore, that if the Dutch government had adopted the Sabin polio vaccine, some children might well have been unwittingly protected against the disease regardless of parental desires. Whatever the parents' religious or philosophical objections, their offspring would have been vaccinated just as effectively as if they had received attenuated poliovirus from a health professional during a routine immunization program.

If that scenario poses a difficult question for medical ethics, so, too, does its logical sequel: that vaccine organisms could be genetically modified specifically to achieve greater dissemination. One laudable aim of such a project might be the need for lower-percentage take-up, compared with a killed vaccine, to achieve an effective level of immunity in the population. However, what of the associated possibility of overriding, by biological imperative, conscientious objections to the protection of children against killing diseases?

References

1. **Feikin, D. R., D. C. Lezotte, R. F. Hamman, D. A. Salmon, R. T. Chen, and R. E. Hoffman.** 2000. Individual and community risks of measles and pertussis associated with personal exemptions to immunization. *JAMA* **284:**3145–3150.

2. **Hanratty, B., T. Holt, E. Duffell, W. Patterson, M. Ramsay, J. M. White, L. Jin, and P. Litton.** 2000. UK measles outbreak in non-immune anthroposophic communities: the implications for the elimination of measles from Europe. *Epidemiol. Infect.* **125:**377–383.

3. **Steiner, R.** 1920. *Physiology and Therapeutics.* Mercury Press, New York, NY.

38 Myxomatosis: Grim Questions

Do we want rabbits to suffer and die from myxomatosis, or should we seek instead to protect them against such a grotesque, disfiguring, painful disease?

In the case of other pathogens and hosts, such questions might appear to be callously superfluous. Myxoma virus, however, poses a genuine dilemma. Though not wholly new, the perplexity was sharpened considerably at the turn of the 21st century by myxomatosis vaccine research in Spain, which threatened pest control in Australia. The conundrum was one of many that ought to attract the attention of all those (like the heir to the British throne) who want to believe in a natural world that is pristine, pleasant, and morally unambiguous.

Genetic modification was the ingredient giving new, tantalizing significance to an infection that was first deliberately introduced into Australia and Europe in the early 1950s. I well recall from that time my own horror when I stumbled over two rabbits dying of the disease as I walked through a riverside field in northern England.

Modern knowledge of this ghastly virus infection, which is characterized by horrible mucilaginous tumors on the skin, dates from just before the turn of the century. In 1895, the government of Uruguay invited Giuseppe Sanarelli, a microbiologist at the University of Siena, to set up an institute of hygiene in Montevideo. Sanarelli accepted, and in the course of establishing the new center, he introduced European domestic rabbits to Uruguay for the first time. They were needed for producing sera containing antibodies against various diseases.

The following year, however, the rabbits developed the extremely infectious and lethal disease, then unknown in Europe, which we now call myxomatosis. Nearly half a century elapsed before the common wild rabbit of Brazil was incriminated as the carrier of the myxoma virus and before the method of transmission, via mosquito bites, was firmly proved. Mosquitoes also convey the disease in Australia and parts of Europe. Rabbit fleas are the major vectors in Britain.

Because the European wild rabbit is a major pest in Australia, agricul-turalists tried repeatedly to establish the myxoma virus there, but they did not succeed until 1950, when infected rabbits were introduced into Murray Valley. The splendid weather that summer and in the following 2 years provided unusually favorable conditions for mosquitoes to breed and travel far. Myxomatosis spread like wildfire. Millions of rabbits, about four-fifths of those in southeastern Australia, succumbed.

Myxomatosis was introduced into France in June 1952, and by the end of the following year it had reached Belgium, Luxembourg, Germany, The Netherlands, and Spain. The first outbreaks in Britain were in the southern county of Kent, and the disease moved so rapidly that by the end of 1955 well over 9/10 of the rabbits in the country were dead.

However, Britain's rabbit population later began to thrive again. Today, the animals are again considered a serious pest and a significant factor in agricultural economics. In Australia, the disease remains important in regulating the rabbit population.

There have, however, been two significant changes since the first efforts at biological control. The first alteration was in the virulence of the myxoma virus. The lethal organism first introduced into Australia killed over 99% of infected rabbits, yet within 12 months, strains had appeared with a mortality rate of only 90%. In subsequent years, even milder strains have prospered, in some cases having mortality rates as low as 20%.

The explanation, as the Australian virologist Macfarlane Burnet observed long ago in *Biological Aspects of Infectious Disease* (1), is simply that there is little point in a microorganism destroying its host in spec-tacular fashion if this leaves it with no prospect of being ferried to other

Animalcules: the Activities, Impacts, and Investigators of Microbes

vulnerable hosts. The chances of further transmission are much greater for a virus that is comparatively mild and which therefore produces a lengthier disease and infectious period than with one that kills its host rapidly.

As Charles Darwin might have predicted, rabbits have changed too. Under the rigorous pressures of regular exposure to a myxoma virus that was highly lethal, resistant strains emerged. During a 7-year period in Australia, resistance increased to such a degree that a virus that originally killed 90% of wild rabbits eventually destroyed only some 30%.

The work by Frank Fenner and his colleagues on the coevolution of rabbits and myxoma viruses in Australia in the 1950s and 1960s is one of Australia's great contributions to microbiology. From that research, two major lessons emerged. First, the conditions for successful microbiological control of a "pest" population are so complex that initial failures need have little significance for the eventual outcome of a project. Second, genetic changes in both the microbial and host populations can quickly alter the ground rules, though the extent to which this happens depends on environmental factors and on intelligent anticipation. Above all, the long-term success of biological control rests on a thorough understanding of the spread and persistence of disease in natural populations.

However, a new dilemma emerged as a result of the development of a new, genetically modified myxoma virus by Juan Torres and his colleagues at the Center for Investigation into Animal Health in Madrid, Spain. Not only was it attenuated, so that it immunized rabbits against myxomatosis, but it also carried a gene coding for a protein from rabbit hemorrhagic disease and thus conferred immunity against that killer, too. The virus (which was tested initially in a population of 300 rabbits on a small island) was disabled so that it could spread from one rabbit to another for one generation only.

The question then, and indeed today, was, Do we really want to protect rabbits against these two lethal infections? However strange the question might appear if applied to other diseases, here it poses genuine difficulty because of the differing perspectives of two different countries.

In Spain, populations of rabbits are sparse. Indeed, Juan Torres claimed that myxomatosis and rabbit hemorrhagic disease were endangering the survival not only of rabbits, but also of their natural predators. In this context, a vaccination initiative could be seen as a conservation measure as important as efforts to protect elephants in Africa or whales in the oceans.

In Australia, on the other hand, rabbits remain serious pests, especially for farmers, and the two infections play an important, "natural" role in restricting their numbers. If the new transgenic myxoma virus were to be released there by mistake, it could jeopardize the success of this biological control. While the organism's limited capacity for replication might seem to limit any danger, further genetic changes could make matters worse.

By ironic coincidence, evidence from work on the microbiological regulation of rabbit populations in Australia provided the sharpest reminder of the risks inherent in such work. In 1995, a calicivirus being considered as an additional agent to control rabbits escaped to the mainland from the island where it was being field tested. Although millions of rabbits died as a result, this all happened before ecological questions (such as possible effects on other species) had been clearly answered.

The moral ambiguity of the challenges posed by a biological control agent such as myxoma virus is one thing. It is an altogether different matter to proceed with such a questionable initiative in an area where serious practical mistakes have already been made.

Reference

1. **Burnet, F. M.** 1940. *Biological Aspects of Infectious Disease.* Cambridge University Press, Cambridge, United Kingdom.

39 Rationalizing Vaccination

As the new millennium unfolds, a new epoch in our defense against communicable disease seems increasingly imminent, too. Obituaries are already being written for the magic bullets that have been the mainstay of antimicrobial warfare over the past half century. At the same time, advances in biotechnology are initiating radical changes in the production and targeting of vaccines. The new prospects for immunization against a greatly extended portfolio of infections, and against cancers, point to a transformation in medicine far more extensive than the "antibiotic revolution" of the past. It may well prove more long lasting, too.

Historians will look back with dismay upon the fecklessness with which we abused powerful antimicrobial drugs. Appreciative of the countless pathogens these weapons have killed, the innumerable lives saved, and the morass of human misery prevented, we have nevertheless deployed them so indiscriminately that we created optimal conditions for the emergence of resistant organisms throughout the world. No doubt antimicrobials will continue to have a place in the physician's armamentarium. Greater prudence in prescribing will help, and researchers will continue to discover new agents and better ways of delivering them, but things can never be the same again.

Drugs and vaccines have, of course, long been complementary approaches in controlling infectious diseases. Today, they contrast starkly. Many antibiotics are declining in potency, just as we are realizing the extreme crudity of an approach that kills commensal organisms, as well as intended targets. Far from being tightly aimed, as Paul Ehrlich believed, magic bullets are actually blunderbusses, now destined to occupy a more restricted niche in medical history.

By comparison, vaccination is a highly specific method of mobilizing the body's own defenses. The craft of vaccine manufacture, built on principles established by Louis Pasteur a century ago, is being transformed by molecular genetics and bioinformatics, and a host of new vaccines, including many effective against diseases not previously susceptible to immunization, are under development for the coming decade.

Even without further innovation, the opportunities are immense. The World Health Organization has estimated that nearly 4 million children still die each year from diseases (such as measles, hepatitis B, rubella, and *Haemophilus influenzae* type b [Hib] infections) that are preventable by existing vaccines. Just one example of an initially underused product is pneumococcal vaccine. As Columbia University public health specialist Jane Sisk and colleagues observed in the *Journal of the American Medical Association* (6), immunization to prevent pneumococcal bacteremia in elderly people is one of the few interventions that both improves patient well-being and saves medical costs. Nevertheless, health authorities in many countries have been slow to make full use of it.

It is, however, the wide variety of emerging vaccines that illustrates the full scale of the transformation likely in the years ahead. One recent estimate is that vaccines against no less than 75 different communicable diseases are currently at various stages of research and development. Samuel Katz of the Duke University School of Medicine has pointed out (4) that only human immunodeficiency virus type 1 and malaria continue to defy the efforts of biotechnologists to prevent infection with new vaccines. Only tuberculosis has resisted attempts to improve an existing, less than optimal product (*Mycobacterium bovis* BCG).

One of the most heartening developments during recent years has been the discovery that certain new vaccines, previously validated in industrialized countries, have also been effective in parts of the world where poor nutrition and other problems might have rendered them less powerful. Thus, trials of a Hib vaccine in The Gambia indicated that it would substantially reduce childhood deaths due to meningitis and pneumonia in developing countries. There have been similar advances with a rotavirus vaccine in Venezuela and a cholera vaccine in Vietnam. One of the most recent successes came with rotavirus vaccine in Latin American children (5).

The Hib vaccine used in The Gambia is an example of a relatively new type of product: conjugate vaccines. These are made by linking an antigenic molecule (in this case, the outer capsular polysaccharide) to a powerful immunogen, but this is only one of a large portfolio of products, from subunit vaccines to DNA vaccines, now being generated by biotechnology.

The ultimate step is the comprehensive identification of all genes associated with virulence and antigenicity through the sequencing of pathogens. Each completed genome will indicate hitherto unrecognized targets for immunization. The earliest returns may well come from *Mycobacterium tuberculosis*, through the incorporation of new proteins in BCG, for example, or the construction of novel attenuated strains by altering base sequences involved in virulence.

Aside from infections, the prospects for immunization against cancers have escalated so rapidly in very recent times that the implications have not yet been widely recognized. Four major examples are carcinoma of the uterine cervix, hepatocellular carcinoma, stomach cancer, and nasopharyngeal carcinoma, which may be preventable by human papillomavirus (HPV), Epstein-Barr virus, *Helicobacter pylori*, and hepatitis B virus vaccines, respectively.

Although there are unresolved social problems in some countries regarding HPV vaccine, it is already showing highly encouraging results against cervical cancer (3). However, while it could possibly save millions of women's lives and is now being given routinely to young girls in the United States and Europe, cost has prevented its deployment in poorer countries, where its impact could be even greater. In 2008, a campaign was launched to introduce HPV vaccine in Latin America.

One early sign of success was the dramatic decline in hepatocellular carcinoma in children since Taiwan established universal immunization against hepatitis B virus in 1984 (1). Evidence from Denmark reported by Morten Frisch and colleagues in Copenhagen (2) has strongly suggested that a sexually transmitted infection also causes anal cancer. The presence of HPV in most anal cancer specimens shows that this, too, is potentially preventable.

Like historic advances in arms control, radical innovations in immunization are unlikely to be free of new challenges and difficulties, above all, how to make optimal use of the enormous catalogue of new vaccines now on the launchpad. Can increasing numbers of potentially lifesaving ingredients be incorporated into infant immunization schedules that are already quite full? Even when measles and polio are deleted, when they have followed smallpox into oblivion, there will be strong cases for the inclusion of many more immunogens than have been used in the past. Will this be acceptable to parents? Will it be acceptable to the human body?

One possible answer is to extend routine immunization beyond the infant years, where it is concentrated at present. Thus, 9- to 12-year-old children might be targeted for immunization against sexually transmitted diseases, such as gonorrhea, which were hitherto not preventable in this way. Protecting young persons before they indulge in high-risk behavior will, however, mean reaching them at an age when they do not normally come into regular contact with health services.

Controlled-release products may also ease the pressure on immunization schedules. Vaccines delivered slowly or intermittently over months or even years could supplant the conventional system of initial doses followed by boosters later in infancy.

Finally, the burgeoning revolution in vaccinology is surely the moment to coordinate our use of these potent prophylactics internationally. It makes little sense, for example, that virtually every member country of the European Union has a different policy or reimbursement schedule for influenza vaccination. Infant immunization varies throughout the world, sometimes reflecting factors such as differences in school entry age but often resulting from insignificant historical accidents. It is time for rationalization.

Animalcules: the Activities, Impacts, and Investigators of Microbes

References

1. **Chang, M.-H., C. J. Chen, M. S. Lai, H. M. Hsu, T. C. Wu, M. S. Kong, D. C. Liang, W. Y. Shau, and D. S. Chen.** 1997. Universal hepatitis B vaccination in Taiwan and the incidence of hepatocellular carcinoma in children. *N. Engl. J. Med.* **336:**1855–1859.

2. **Frisch, M., B. Glimelius, A. J. C. van den Brule, J. Wohlfahrt, C. J. L. M. Meijer, J. M. M. Walboomers, S. Goldman, C. Svensson, H.-O. Adami, and M. Melbye.** 1997. Sexually transmitted infection as a cause of anal cancer. *N. Engl. J. Med.* **337:**1350–1358.

3. **Kahn, J. A., and R. D. Burk.** 2007. Papillomavirus vaccines in perspective. *Lancet* **369:**2135–2137.

4. **Katz, S. L.** 1997. Future vaccines and a global perspective. *Lancet* **350:**1767–1770.

5. **Linhares, A. C., F. R. Velázquez, I. Pérez-Schael, X. Sáez-Llorens, H. Abate, F. Espinoza, P. López, M. Macías-Parra, E. Ortega-Barría, D. M. Rivera-Medina, L. Rivera, N. Pavía-Ruz, E. Nuñez, S. Damaso, G. M. Ruiz-Palacios, B. De Vos, M. O'Ryan, P. Gillard, A. Bouckenooghe, and the Human Rotavirus Vaccine Study Group.** 2008. Efficacy and safety of an oral live attenuated human rotavirus vaccine against rotavirus gastroenteritis during the first 2 years of life in Latin American infants: a randomised, double-blind, placebo-controlled phase III study. *Lancet* **371:**1181–1189.

6. **Sisk, J. E., A. J. Moskowitz, W. Whang, J. D. Lin, D. S. Fedson, A. M. McBean, J. F. Plouffe, M. S. Cetron, and J. C. Butler.** 1997. Cost-effectiveness of vaccination against pneumococcal bacteremia among elderly people. *JAMA* **278:**1333–1339.

40 A European Furor

"Whenever I visit Europe these days, I find it hard to understand what is going on" was the comment I heard some years ago from an official prominent in the U.S. biotechnology industry. He was referring to the very different social, regulatory, and political climates surrounding biotechnology on opposite sides of the Atlantic. Genetically modified (GM) crops have been the prime example in recent years. Though cultivated and widely accepted in the United States, they attracted public hostility in most European countries, where various types of moratoria were instituted.

So what was going on? The debate was complex. However, close examination reveals that much of the apparent public antipathy to GM foods has stemmed from concerns that were primarily microbiological, rather than botanical.

There are lessons here for other countries, at other times, which may encounter opposition to scientific developments from directions as unexpected as they are irrelevant. When this does happen, incoherent frustration of the sort shown recently by certain individuals and organizations in Europe is not a helpful response.

Outstanding among the sources of discontent—sometimes specifically cited, sometimes simply a ghostly presence at the debate—was bovine spongiform encephalopathy (BSE). Through an accident of history, the genetic engineering of foods was contemporaneous with the recognition of BSE in United Kingdom cattle and the subsequent emergence of "new-variant" Creutzfeldt-Jakob disease (CJD) in young people in 1996. These unquestionably serious developments precipitated a massive cull of cows in Britain and triggered alarms in continental countries and bans on the importation of British beef.

However, those events had nothing whatever to do with plant breeding, ancient or modern. New-variant CJD almost certainly followed the consumption of tissue from cattle infected with BSE, which had, in turn, eaten meat or bone meal contaminated with rendered carcasses of sheep infected with scrapie. The root cause was probably a change in the animal-rendering process that allowed the scrapie agent to survive.

However, it was the BSE affair that was endlessly quoted, in newspaper articles and on television and radio programs, as the basis upon which GM foods should be rejected. "Bland assurances from governments and agribusiness don't wash any more, not after BSE," said the United Kingdom's highly regarded and influential *Good Food Guide* (1). "If BSE has taught us anything, it is surely to be cautious about tampering with natural processes, however well-intentioned, however plausibly the benefits are packaged."

Speaking on television to justify his company's ban on GM ingredients in its own-brand products, the chief executive of Iceland Supermarkets, Malcolm Walker, gave BSE and interference with the natural world as the grounds for their action. "We are corrupting nature and we have sufficient evidence to show that nature fights back—salmonella, listeria, BSE," he wrote in *The Times* newspaper (2).

The most extraordinary feature of the use of BSE to damn genetic manipulation was that the one significant link between these two domains points to a very different conclusion. In the real world, production of human growth hormone in bacteria eradicated the problem of CJD caused by the hormone formerly extracted from the pituitary glands of human cadavers. Moreover, BSE would in all probability never have emerged at all if the rendering of animal carcasses, instead of being deregulated 20 years ago by the government of Prime Minister

Margaret Thatcher, had been as strictly controlled as is plant biotechnology today.

However, BSE was by no means the only microbial argument brandished in the crusade against GM foodstuffs. "Use of genetically modified bacteria in the food supplement tryptophan may have caused 37 deaths in the USA since 1989 as well as permanently disabling thousands of people," said an Iceland leaflet explaining its ban to customers. "Once released into the natural environment, genetically modified bacteria and plants interbreed with those in the wild. . . . Some scientists fear the development of antibiotic-resistant 'superbacteria.'"

Misapprehensions based on the microbiological theme multiplied in widely reported statements from some of Britain's most distinguished chefs, in support of efforts by Friends of the Earth to halt the sale of GM foods. The head chef at one prestigious London restaurant, The Square, said that only now were we realizing that antibiotics had "trashed the flavor of what we eat."

Such breathtaking ignorance recalls the period in the mid-1980s when, during parliamentary and media discussions of animal feedstuffs, participants frequently confused antibiotics with hormones. In despair, the British Veterinary Association issued a briefing document explaining the difference.

There are, of course, legitimate safety questions surrounding the development of GM crops, in particular ecological issues. For example, will pollen grains ferry transgenes into non-GM plants, where they may have unforeseen consequences? Fortunately, there is a wealth of experience from the past to help in answering questions of this sort.

One such case is the cultivation of oilseed rape for human consumption, which contains low levels of erucic acid, and of industrial oilseed rape, whose high erucic acid content makes it toxic to humans. These are simply grown far enough apart to prevent cross-pollination. In the case of novel crops, precautions of this sort are being determined on the basis of field trials and risk assessment studies. In our present state of knowledge, it would seem essential, for example, to prevent genes coding for the production of vaccine proteins in plants from entering the general food chain.

Ironically, European campaigners against GM foods adopted as one of their principal tactics the disruption of field trials that were specifically established to evaluate the safety of GM crops. After 1996, when *Nature* reported that every one of 15 trials conducted by universities and research institutes in Germany had been attacked, this type of conduct became commonplace in other countries, especially the United Kingdom.

There is a perplexing discontinuity between the environmental consequences of cultivating transgenic crops, which rightly attract considerable scientific interest, and the issues repeatedly raised in public discourse, whether by campaigning groups, supermarket chiefs, or celebrity cooks. Paramount among these were the alleged dangers of consuming such plants. These were founded not only upon the BSE affair, but also on a whole catalogue of natural pathogens, from *Salmonella* and *Listeria* to *Escherichia coli* O157, and the outbreaks they have caused.

How did Europeans get into such a mess? Two observations seem pertinent. First, because scientists are (or try to be) rational creatures, they have an instinctive trust that rational solutions will prevail. Faced with assertions about transgenic plants based on BSE and about antibiotics destroying the flavor of food, they recoil in dismay, yet they believe that the absurdity of these claims will be quickly recognized. In not bothering to respond to allegations of this sort, they may well have felt supported by evidence from opinion polls, such as the Eurobarometer. These showed keen public awareness of the contributions that science and technology make to the quality of life.

Secondly, both the unreality of many attacks on genetic manipulation and scientists' professional inclinations inhibited them from going on the offensive. I have yet to hear, for example, a single advocate of transgenic-plant research demolishing the simplistic argument that we should not "interfere with nature." Far from being the wholesome, pristine condition imagined by its healthy, well-nourished Western protagonists, nature was a world of pestilence, famine, and horrendous plagues; smallpox, diphtheria, and paralytic poliomyelitis; and a human life expectancy of 40 years.

The European furor over GM foods has many ingredients. They include historical concerns about genetics and eugenics, especially in Germany; suspicion of large companies; unease over intensive agriculture; and, in the United Kingdom, ex cathedra pronouncements by the heir to the throne. In retrospect, however, the scientific establishment's failure to combat microbiological nonsense may prove to have been equally significant, and regrettable.

References

1. **Ainsworth, J.** 1999. *The Good Food Guide 1999.* Consumers Association, London, United Kingdom.

2. **Walker, M. C.** 24 January 1997. *The Times,* London, United Kingdom.

41 Bioremediation and Greenery

Imagine that you are an industrialist charged with the development of a particular technology. You are conscious of your responsibilities, not only to your staff and shareholders, but also to your fellow citizens and to the wider world. Now consider the three most desirable criteria to which you would like the technology to conform, in addition to the self-evident need to be useful and, indeed, profitable.

The answer is likely to be as follows. The technology should be ecologically acceptable, neither polluting the biosphere nor consuming finite resources, and ideally actually improving the quality of the natural environment. Secondly, it should not be radically new, because this would bring inherent uncertainties as to its consequences. Thirdly, it should, for the first two reasons, be a technology that can be confidently commended to the public and to politicians and regulators.

Bioremediation—using microorganisms to degrade environmental pollutants—is surely such a technology. Being "green" in both purpose and methodology, it meets my first criterion as an activity whose only impact on the environment is through its cleansing role. Bioremediation is appropriate, too, for application in both developing and developed countries.

Moreover, as bioremediation is not intrinsically novel, it meets my second criterion. The transforming roles played by microorganisms when used to attack pollutants in the soil and elsewhere are comparable with those of bacteria and other organisms involved in the worldwide process of sewage disposal. It is difficult to draw any clear line between dedicated, scientifically based bioremediation and, for example, the continual recycling of elements in the biosphere or the sort of natural cleansing that occurred along the Kuwaiti and Saudi Arabian coast following the deliberate leaking of oil during the occupation of Kuwait in 1990 and 1991.

Thus, bioremediation is neither unnatural nor inherently novel. This, in turn, indicates that its practitioners are at least partially free of the criticism, which can be leveled against any genuinely new technology, that its consequences and dangers are unpredictable.

Public acceptance, my third criterion, is by no means guaranteed by fulfillment of the other two criteria. Nevertheless, there is an enormous difference between the conscious extension of organic processes that fashion the biosphere day by day, century by century, and the introduction of a technology founded upon approaches that mark a sharp departure from that which has gone on before.

What exactly is likely to be the public mood concerning the emerging technology of bioremediation, then? Some reports might suggest a very gloomy answer, certainly as regards the use of genetically manipulated organisms for this purpose, yet surveys conducted in both the United States and Europe do not support the idea that there is a high level of public opposition to biotechnology or genetic engineering. They do, however, indicate that it is simplistic and misleading to measure the temperature of public opinion along a single scale from enthusiasm to hostility.

In one United Kingdom study (1), the investigators measured general attitudes in the community by asking nine questions, ranging from whether scientists can be trusted to whether science is changing our way of life too quickly. The results showed a generally positive view of science, with 70% of respondents believing that science and technology are making our lives healthier and more comfortable and 80% backing government support for "research which advances the frontiers of knowledge." However, the pattern was not uniform. Some participants agreed with both positive and negative statements.

Animalcules: the Activities, Impacts, and Investigators of Microbes

To determine whether the subjects' overall view of science was reflected in their opinions on individual areas of research, Evans and Durant used further sets of questions. The specific topics included creating new forms of animal life and finding a cure for cancer. Here, the results revealed a correlation that was at best moderate, being best for useful and basic science and weakest for morally contentious research. In other words, participants' responses to questions exploring their general attitudes toward science and technology did not permit accurate predictions of how they felt about particular types of research.

A third analysis compared the respondents' attitudes to science with their factual knowledge, as assessed from their responses to a diversity of statements from the natural and medical sciences, for example, that light travels faster than sound and that sunlight causes skin cancer. Here, Evans and Durant discovered that although there were statistically significant correlations between attitude and knowledge, the strength of the link varied considerably.

Factual knowledge did correlate moderately well with both the participants' attitudes in general and those toward useful and basic research. On the other hand, knowledge was almost wholly unrelated to attitudes to nonuseful research, and there was a strong negative association between knowledge and support for research that can be seen as morally contentious.

Other studies, which have invited comments on specific applications of biotechnology, provide more concrete evidence that helps us to predict the likely public mood in regard to the use of microorganisms in bioremediation. Consider, for example, a survey on attitudes to biotechnology carried out in the United Kingdom at a time when there was considerable controversy, and certainly media consternation, over gene technology (2). The researchers evaluated attitudes among a population sample that included people, chosen at random from electoral registers, living in two parts of the country where releases of genetically manipulated organisms had occurred; members of Friends of the Earth and other "green" groups; and nontechnical staff of companies involved in biotechnology.

Even with a sample of this composition, which might have been expected to bias the results away from positive endorsement of the technology, support for the use of genetic manipulation for environmental decontamination was as strong as for its use in medicine. Some 65% of respondents were "comfortable" with the idea of deploying recombinant organisms to clean up oil slicks and to detoxify industrial waste. This compared with 59% in the case of medical research and 57% in the case of "making medicines."

Thus, the hard evidence does not seem to justify the worst fears of many commentators as regards alleged public hostility toward biotechnology, and of the various applications of biotechnology, bioremediation appears to be one of the most likely to enjoy public confidence and support.

A different type of argument to bolster this conclusion comes from the part of Europe that experienced the most vigorous antagonism toward gene technology, the former West Germany. Such opposition has abated over the past 20 years, and in *Resistance to New Technology*, Joachim Radkau of the University of Bielefeld has attributed the beginning of this change in part to alterations in Germany associated with reunification. Not only did the economic and social problems arising from reunification divert attention away from the alleged dangers of gene technology, in addition, allegations concerning these risks and hazards were suddenly overshadowed as people learned more about the scale of industrial pollution in East Germany and, indeed, other eastern states. "The risks of genetic engineering seemed relatively harmless beside the destroyed landscape resulting from East Germany chemistry and brown coal," Radkau wrote (4).

Writing in 2005 on the basis of a survey of public opinion carried out in the European Union, the United States, and Canada, George Gaskell of the London School of Economics and Political Science pointed out that Europeans had become much more positive about biotechnology (2). People seemingly associated biotechnology with the human genome project and medical applications, which they saw in a positive light. "The public expect and want science and technology to solve problems, but they also want a say in deciding which problems are worth solving," Gaskell wrote. "This is not a matter of attracting public support for an agenda already established by science and scientists, but rather of seeing the public as participants in science policy with whom a shared vision of socially viable science and technological innovation can be achieved."

Animalcules: the Activities, Impacts, and Investigators of Microbes

Opinions vary as to why bioremediation is not being more widely and more vigorously adopted. However, there is little doubt that nervousness concerning public acceptability is at least one significant component behind the current, somewhat hesitant mood. However, despite the fact that more information, better understood, does not guarantee public acceptance, this is a technology that is as well positioned for public endorsement as any technology is ever likely to be again.

References

1. **Evans, G., and J. Durant.** 1995. The relationship between knowledge and attitudes in the public understanding of science in Britain. *Public Understanding Sci.* **4:**57–74.

2. **Gaskell, G., E. Einsiedel, W. Hallman, S. H. Priest, J. Jackson, and J. Olsthoorn.** 2005. Social values and the governance of science. *Science* **310:**1908–1909.

3. **Martin, S., and J. Tait.** 1992. *Science in Public.* Science Museum, London, United Kingdom.

4. **Radkau, J.** 1995. Learning from Chernobyl for the fight against genetics? Stages and stimuli of German protest movements—a comparative analysis, p. 335–355. *In* M. Bauer (ed.), *Resistance to New Technology.* Cambridge University Press, Cambridge, United Kingdom.

42 The Citation Game

Citation analysis provokes strong but mixed reactions among most working scientists. Some have, consciously or otherwise, benefited from career rewards associated with publication in high-impact journals. Some have been appointed or promoted, at least in part, on their scientometric record. Others have failed to hit the jackpot and are thus frustrated over what they see as distortions introduced into the process of science by overreliance on citation ratings.

Odd researchers will bend your ear about malpractices (mutual citation among friends, for example) that are in reality very rare. I once knew a bacteriologist who believed quite wrongly that his professional advancement had been retarded because his surname began with Y rather than A or B. Another complained, after retirement, that his successor as head of a prestigious lab was failing to credit him in published papers. When he was shown that this was untrue, he protested instead that he was being cited without permission.

The *Science Citation Index* was not, of course, originally created so that editors, administrators, and grant-awarding committees could pore over citation ratings and use them to compare the merits of one individual, institution, or country with those of another. League tables and "citation classics" (such as Alick Isaacs and Jean Lindenmann's first

Animalcules: the Activities, Impacts, and Investigators of Microbes

interferon report [see chapter 50]) were not on the original agenda for Eugene Garfield, who 40 years ago conceived the *Science Citation Index* simply as a tool for bibliographical retrieval.

Over the last 2 decades, however, citation counts have assumed considerable significance in science policy and in the evaluation of scientific performance. While Garfield himself was the first to urge caution, others have seized upon scientometric data as offering a solid, quantitative basis upon which to support research, make appointments, and target journals for the submission of papers.

However, is the central notion of ranking citations valid as a criterion of quality in science? Clearly, a paper that quickly becomes highly cited has made a greater impact than one that is totally ignored (they do exist). Moreover, a scientist who repeatedly scores relatively high over the years is having a major influence in shaping the development of his or her field.

However, there are important caveats, too. One stems from the concept of the "sleeper," the paper that is barely recognized for many years until its relevance suddenly becomes apparent and citations accrue accordingly. If the literature is analyzed over too short a time interval, such key papers will be missed.

What is surprising is that citation analysts have not gone much beyond their quantitative data to ask deeper questions about the process and, indeed, the politics of science. True, there have been thematic mutations, such as cocitation analysis. By focusing on papers that cite other papers in common, this approach has usefully highlighted the strength of intellectual links between publications in disparate disciplines. A case in point is the advent of psychoimmunology about 25 years ago, presaged by increasing cocitation between the immunology and psychology literatures.

The outbreak of severe acute respiratory syndrome (SARS) in 2002–2003 provided an opportunity for citation analysts to make a far more potent point, one that could have been influential in terms of science policy. Although the Chinese authorities appeared to be slow to inform the rest of the world about SARS when it emerged in November 2002, later progress in understanding the condition was astonishingly quick.

The World Health Organization issued its first global warning on 12 March 2003. Less than a month later, on 8 April, a very detailed paper about SARS and the coronavirus thought to be responsible for the syndrome, written by researchers from Hong Kong, appeared in the online edition of *The Lancet*, and on 12 April, the British Columbia Cancer Research Centre released its complete genome data on the organism.

"The pace of SARS research has been astounding," said David Heymann of the World Health Organization. "Because of an extraordinary collaboration among laboratories from countries around the world, we now know with certainty what causes SARS.... Now we can move away from methods like isolation and quarantine and move aggressively towards modern intervention strategies including specific treatments and eventually vaccination."

Leaving aside the question of what happened to Koch's postulates in this instance, there was surely a golden opportunity here for citation analysts to demonstrate why progress had been so rapid. Heymann was right about international collaboration, but that was not the whole story. Even a cursory examination of the first few SARS reports, particularly the papers they cited, showed that investigators were able to move so quickly only by using existing knowledge and techniques available to them.

The Hong Kong communication, published in *The Lancet* (1), carried citations to earlier papers on, inter alia, tests for the rapid identification of influenza A and B virus infections, the persistence of human coronaviruses OC43 and 229E in neural cell cultures, and the induction of proinflammatory cytokines in human macrophages by influenza A viruses. Another described the use of PCR to amplify conserved sequences in the matrix gene of influenza A virus.

All of these references may seem unremarkable: they simply reflected the normal, incremental story of advancing science. From the point of view of research funding, however, there is a powerful case to be made here to illustrate the importance of fundamental studies whose findings may be inordinately helpful when we are confronted with a crisis such as that triggered by SARS.

The same was true with the advent of AIDS and human immunodeficiency virus (HIV). Looking back to those early papers, one is reminded of the vast infrastructure of microbiology that facilitated progress in characterizing an infection even more baffling than SARS. There were references to T cells and other aspects of the immune response, to sequencing methodology, retroviruses, and many other topics brought together in an effort to characterize AIDS and its virus.

None of this work could, of course, have been conducted specifically to throw light on the new arrivals HIV and AIDS. It had been carried out, and funded, for other purposes. Some of the work, on the immune system, had been supported for no specific practical purpose whatever but simply as basic science, yet the results were there, years later, when they were urgently required.

Back in 1971, United Kingdom science was thrown into turmoil by the so-called Rothschild report, which appeared to threaten the funding of fundamental science. In response, countless scientists wrote to *The Times* to remind the government of the importance of pure research as the seedbed from which important applications later flowed. Charles Dodds, for example, pointed out that his 1930s work on the hormones regulating fertility in mammals had led subsequently to the synthesis of diethylstilbestrol, later used to treat prostate cancer.

How much more compelling this case would have been if it had been supported by citation studies highlighting Dodds' papers and the earlier work they cited, together with information about the sources of funding in each case and the reasons for which it was given. A bonus would have been information about the practical and commercial returns from diethylstilbestrol.

Likewise 3 decades later, SARS and HIV/AIDS, because of both their inherent novelty and the novel challenges they posed, would have been excellent candidates for such scientometric research. The outcome would no doubt be a vivid demonstration of our dependence, when confronted by emerging infections, upon research conducted for quite different purposes (and in some instances for no practical purpose whatever) years before.

It is notoriously difficult to capture the attention of politicians regarding topics that are not of obvious and immediate concern. This type of analysis might just do the trick.

Reference

1. **Peiris, J. S. M., S. T. Lai, L. L. Poon, Y. Guan, L. Y. Yam, W. Lim, J. Nicholls, W. K. Yee, W. W. Yan, M. T. Cheung, V. C. Cheng, K. H. Chan, D. N. Tsang, R. W. Yung, T. K. Ng, K. Y. Yuen, and the SARS Study Group.** 2003. Coronavirus as a possible cause of severe acute respiratory syndrome. *Lancet* **361:**1319–1325.

IV. Personalia

43 Antony van Leeuwenhoek, Clifford Dobell, and Robert Hooke

Rarely in the history of science has there been such a perfect match of subject and biographer as that between the pioneer microscopist Antony van Leeuwenhoek and Clifford Dobell, his devoted chronicler and translator. Dobell's book, *Antony van Leeuwenhoek and His "Little Animals"* (2), published in 1932 after 25 years of dedicated research, is uncannily reminiscent of the meticulous reports through which Leeuwenhoek communicated his discoveries of "animalcules" in the late 17th and early 18th centuries.

Thanks almost entirely to Dobell, the Dutchman's story, an incredible one by any standards, is now known in detail. Leeuwenhoek was a draper, with no scientific training, who constructed such excellent (though simple) microscopes, and used them so skillfully, that he was able to see bacteria, protozoa, spermatozoa, blood corpuscles, and many other objects invisible to the naked eye.

Initially with diffidence, but later with ebullient enthusiasm, Leeuwenhoek reported his discoveries, made assiduously by scrutinizing everything from rainwater and frog's skin to his own semen and scrapings from his teeth, in a series of letters to the Royal Society in London. Some 120 extracts of these appeared in the *Philosophical Transactions*, and in 1680, this remarkable man, who had been apprenticed to a linen draper at the age of 16, was elected a Fellow of the Royal Society.

Every one of the Dutch pioneer's microscopes consisted of a single, very small, biconvex magnifying glass, which he mounted between small apertures in two thin oblong metal (usually brass) plates riveted together. The object to be examined was mounted on a silver needle on the other side of the lens, with glass capillary tubes used to hold liquids that might contain items of interest. The user held the instrument close to the eye and used little thumb screws to adjust the focus.

It was Regnier de Graaf (the man who discovered the Graafian follicle in the ovary) who introduced Leeuwenhoek to the Royal Society. Writing from their hometown of Delft in The Netherlands, he described his friend as "a certain most ingenious person here." Though the response was positive, Leeuwenhoek's initial letter was abject in the extreme. He had, he explained, declined earlier requests to record his observations on paper "because I have no style, or pen, wherewith to express my thoughts properly" and "because I have not been brought up to languages or arts, only to business." He now saw that his "observations did not displease the Royal Society" and prayed that the Fellows "take not amiss my poor pen, and the liberty I here take in setting down my random notions."

Leeuwenhoek soon forgot such humility, as he was carried away by the never-ending task of exploring the microbial world, which he was well aware of revealing to human senses for the first time. Much of his time, too, was spent in grinding lenses, which had exceptional clarity for their time, and in improving his microscopes. When he died in 1723, he left 247 completed microscopes and 172 lenses. What he never divulged was his method of using them, but the excellent results he achieved with his instruments, which were extremely primitive by modern standards, suggest that he did in fact discover the principle of dark-ground illumination.

The affinity of Clifford Dobell for his subject can be seen by perusing the "epistle to the reader" that prefaces his book. It has the same affectionate, personal style as Leeuwenhoek's own letters. The epistle contains similar flourishes of both humility and arrogance. Above all, it reflects a comparable amount of dogged research and scrupulous scholarship.

Dobell was himself a microbiologist and became interested in Leeuwenhoek when, first as a youngster peering at organisms in hay infusions and then as a researcher studying protozoa in frogs, and later

Animalcules: the Activities, Impacts, and Investigators of Microbes

again when investigating oral spirochetes and other bacteria, he found that all microbiology seemed to originate with the Dutch lens grinder from Delft. However, very little was written about this mysterious figure, and such books as were available were hopelessly contradictory. "No two writers gave the same account of him, even when copying one another," Dobell recorded in disgust.

Dobell set out to do his own historical research. First, he discovered that the only English version of Leeuwenhoek's letters was badly garbled (and all references to spermatozoa, and other passages "which to many readers might be offensive" had been deleted). He decided, therefore, to study the original Latin texts. These proved to be "not written in the Latin which I had learned at school" but written in a weird "Leeuwen-hoekian Latin" that was actually nearer to Dutch.

Dobell turned instead, therefore, to a Dutch edition of the works and tried to learn that language. After plowing through the Dutch letters, using the Latin versions and an old Dutch bible for cross-checking, he gained a smattering of 17th-century Latin and Dutch. Then he heard that the original manuscripts were available at the Royal Society in London and went to consult them, only to find that he could not read a single word of the strange script they contained.

This was in the midst of World War I, and being otherwise occupied, Dobell gave up the task. When peace came, he returned to the Royal Society library and picked up the threads. Gradually, he managed to discern words, sentences, and eventually entire passages, a labor of love he felt must be nearly as great as that which had faced those who unraveled the hieroglyphs of the Rosetta Stone.

"You may laugh, dear Reader, but it is true," Dobell wrote. "You, who are doubtless familiar with Dutch and Latin and English of all periods, and for whom the deciphering of ancient manuscripts holds no terrors, must please try to put yourself in my ignorant position and consider my handicaps. . . . I had nobody to help me, and therefore made the most pitiable mistakes."

Only after completing the initial, laborious work of translation could Dobell embark on his second task, a massive historical search for every paper and manuscript that could cast further light on Leeuwenhoek and his researches. Gradually, during a quarter century of devoted study, a picture of the Dutchman's work and personality began to emerge. That picture, as recorded by Dobell, is a compelling one. It is in no way spoiled by the author's gentle but firm reminders, like those of Leeuwenhoek himself, that what he has accomplished is rather splendid.

The oft-repeated assertion that Leeuwenhoek was the "first microbiologist" has been strongly challenged in recent years. In particular, Howard Gest has pointed out that another Fellow of the Royal Society, Robert Hooke, presented the first depiction of a microorganism (the microfungus *Mucor*) in his *Micrographia*, published in 1665. Like the Dutchman, the extraordinarily polymathic English scientist Hooke was able to make his observations because of his ingenuity in fabricating and using simple microscopes.

Writing in the *Notes and Records of the Royal Society* (3), Gest notes that, despite its scholarship, Dobell's book omitted any mention of Hooke's discovery of microfungi and inexplicably dismissed Hooke's great contribution in a footnote (which oddly stated that "it is impossible and unnecessary to discuss this remarkable man and his work here"). The failure of Dobell's index to cite *Micrographia* is certainly bizarre. Gest is also right to point out (4) that Paul de Kruif's astronomically popular *Microbe Hunters* (1) overdramatized the story of Leeuwenhoek and thereby diminished that of Hooke.

Doubtless history will record that microorganisms were discovered by "two remarkable geniuses," as Gest calls them, Antony van Leeuwenhoek and Robert Hooke. Both were extraordinarily perceptive, in both meanings of that word, and technically gifted, given the simplicity of their microscopes. Perversely, perhaps, I still feel that the range of microorganisms discerned by Leeuwenhoek gives him a strong case for being remembered as the first microbiologist—but then, he did have an unusually persuasive biographer.

References

1. **de Kruif, P.** 1926. *Microbe Hunters.* Harcourt, Brace, and Co., New York, NY.

2. **Dobell, C.** 1932. *Antony van Leeuwenhoek and His "Little Animals."* Harcourt, Brace, and Co., New York, NY.

3. **Gest, H.** 2004. The discovery of microorganisms by Robert Hooke and Antoni van Leeuwenhoek, Fellows of The Royal Society. *Notes Rec. R. Soc.* **58:**187–201.

4. **Gest, H.** 2007. Fresh views of 17th-century discoveries by Hooke and van Leeuwenhoek. *Microbe* **2:**483–488.

44 Robert Koch
and His Postulates

Specific etiology was one of the most powerful and productive ideas in the entire history of medicine. Replacing a shadowy intellectual landscape of humors and miasmata with the notion that particular diseases had particular causes, it marked the beginning of the end for both diagnostic imprecision and clinical impotence.

Within the domain we now recognize as microbiology, the prime mover was, of course, Louis Pasteur. He had the genius to discern that the characteristic presence of distinctive types of microbes in different "diseases" of beer and other natural processes might be paralleled in human maladies. However, it was Robert Koch who outlined the postulates according to which an organism could be conclusively implicated as the agent of a specific disease.

The now-familiar form of the four postulates closely followed Koch's own work on anthrax. An organism under suspicion as a disease agent should be present in all cases of the disease. It should be cultured in pure form in the laboratory, cause the same disease when inoculated into a healthy animal, and be isolated again from the lesions of the disease. There have, in fact, been significantly different versions of these tenets over the years, as observed by K. Codell Carter (1). He also claimed that Edwin Klebs, rather than Jacob Henle, should be credited alongside Koch as their creator.

Animalcules: the Activities, Impacts, and Investigators of Microbes

Whatever the historical truth, and however useful they have undoubtedly been, the postulates have also spawned many problems. René Dubos highlighted one such in *Mirage of Health* (3) while discussing Koch's work on tuberculosis. "Most of the persons present in the very room where he read his epoch-making paper in 1882 had been at some time infected with tubercle bacilli and probably still carried virulent infection in their bodies," he wrote. "At that time, in Europe, practically all city dwellers were infected, even though only a relatively small percentage of them developed tuberculosis." In a population where *Mycobacterium tuberculosis* infection is universal, Dubos suggested, the real cause of the disease is not the bacterium but the malnutrition and exhausting work that, for some unfortunate persons, convert infection into pathology.

Half a century later, there are additional grounds for asking whether the tenets of specific etiology can be considered realistic any more. As well as more sophisticated comprehension of the ecological context explored by Dubos, our understanding of pathogenesis is being continually modified by insights from disciplines such as microbial population genetics and molecular ecology.

We now recognize conditions triggered, not by single, defined microorganisms, but by consortia coexisting in biofilms. Another type of difficulty is posed by organisms that cannot be cultivated in the laboratory. A third stems from our considerably wider perspective on a disease such as cholera, whose epidemics cannot be fathomed on the traditional, simplistic model of pathogen and host, but only through analysis ranging over fields as diverse as climatic change and human social behavior.

Among several attempts over the years to brush up Koch's postulates to take account of new knowledge, Stanley Falkow's paper (4) was an elegant restatement in light of modern molecular genetics. Though conceptually updated, his analysis was notable in continuing to offer the postulates as practical tools.

Less successful have been efforts to redraft the rules in the context of present-day ecological analyses of communicable disease. One recent proposal came from Timothy Inglis of the University of Western Australia in Nedlands, Australia. His purpose, as explained in the *Journal of Medical Microbiology* (5), was to take account not only of advances in molecular genetics and ecology, but also of developments in scientific method and the philosophy of science. These included "growing ethical objection to the use of laboratory animal models for incremental scientific gain," which "places a restraint on the use of animal models for pathogenesis research or clinical diagnostic work."

Conscious of the wider range of strategies now used to build an argument for a causal relationship, Inglis proposed what he called a more inclusive approach to establish proof of causality. He built up his argument from a series of assertions. The first was "congruence or reproducible correlation of a taxonomically defined life form with the clinic-pathological and epidemiological features of infection." The second was "consistency of the demonstrable biological response in the subject to an encounter with the prospective infective agent." The third was "progressive or cumulative dissonance as an explanation for pathophysiological processes at every known level of biological organisation in the subject." The last was "curtailment of that pathophysiological process on the deliberate introduction of a specified biomedical intervention."

Inglis asked us to recognize a subcategory of microorganisms to be called "priobes" based on "evidence to implicate the candidate biological entity as an initiator or primer for cumulative dissonance." A priobe is "the sufficient and necessary antecedent cause of a pathophysiological process evident as an infectious disease."

I have studied carefully, and wrestled with, these notions. I have even found them conceptually enticing. However, Koch's postulates were not devised as intellectual diversions. They were intended, and indeed served robustly for some years, as explicit guidelines for research. Having tried to apply Inglis' proposals to several currently unresolved issues in disease etiology, I find it hard to see where they take us.

Can they, for instance, be of any practical help to the research group in Venezuela that has been battling to understand the significance of *Helicobacter*, already familiar for its association with gastritis and gastric ulcers in humans, in Thoroughbred racehorses? The background to their work is a clutch of recent findings: isolation of a new enterohepatic *Helicobacter* species from two healthy horses, a report on a significantly raised prevalence of *Helicobacter equorum* DNA in hospitalized horses, and studies demonstrating a high incidence of gastric ulcers in racehorses in training.

Animalcules: the Activities, Impacts, and Investigators of Microbes

The aim of the new research was to assess the presence of *Helicobacter* DNA in the gastric mucosa of 20 Thoroughbred horses in Caracas, Venezuela. None of the animals had shown any symptoms of gastrointestinal disease. The results of PCR tests on squamous and glandular mucosa samples (2) showed that seven of the horses had gastric ulceration, five had gastritis, and six had both conditions. *Helicobacter*-like DNA was evident in two of the horses with gastric ulceration, in three of those with gastritis, in five of those with both pathologies, and in one with normal gastric mucosa.

Overall, therefore, 10 of the 11 infected animals showed gastric lesions, with just 1 of them having normal mucosa. "This suggests that *Helicobacter* species are present in the stomach of Thoroughbred horses," the investigators conclude. "However, 39% of our horses with gastric pathologies did not show *Helicobacter* or other bacteria, indicating that lesions also may be due to other causes. . . . This is the first report of *Helicobacter*-like DNA in the gastric mucosa of horses. The pathogenic potential of these organisms requires further investigation."

I struggle to see how Inglis' proposals could have helped to clarify this picture, or indeed could help to shape the next phase of research. Still, once we start pondering the issue, we are confronted with a wider question. In an age of molecular genetics, viable but not culturable organisms, biofilms, polymicrobial communities, transmissible spongiform encephalopathies, and highly sophisticated microbial ecology, is there a place for Koch's postulates at all?

References

1. **Carter, K. C.** 1985. Koch's postulates in relation to the work of Jacob Henle and Edwin Klebs. *Med. Hist.* **29:**353–374.

2. **Contreras, M., A. Morales, M. A. García-Amado, M. De Vera, V. Bermúdez, and P. Gueneau.** 2007. Detection of *Helicobacter*-like DNA in the gastric mucosa of Thoroughbred horses. *Lett. Appl. Microbiol.* **45:**553–557.

3. **Dubos, R.** 1960. *Mirage of Health*. Allen & Unwin, London, United Kingdom.

4. **Falkow, S.** 1988. Molecular Koch's postulates applied to microbial pathogenicity. *Rev. Infect. Dis.* **10**(Suppl. 2):S274–S276.

5. **Inglis, T. J.** 2007. Principia aetiologica: taking causality beyond Koch's postulates. *J. Med. Microbiol.* **56:**1419–1422.

45 Hideyo Noguchi, Max Theiler, and Yellowjack

A century ago, the Japanese bacteriologist Hideyo Noguchi became a member of the staff of the Rockefeller Institute in New York, a post he held until 1928, when he died at the age of 51. It is an appropriate time, then, to reflect on the career of this charming, dedicated, yet obstinate man and on how it intersected with that of his contemporary, South Africa-born Max Theiler.

Theiler achieved international acclaim and a Nobel Prize in 1951 after evolving one of the most successful vaccines ever developed, the 17D vaccine against yellow fever. Noguchi, despite other substantial attainments, strayed onto a false trail in pursuit of the organism responsible for the same disease. He even succumbed to yellow fever himself, committing, some have alleged, a microbiologist's hara-kiri once he suspected the truth about his failure.

Theiler first, though. He was born in 1899 on a farm near Pretoria, the youngest child of veterinary bacteriologist Arnold Theiler. As a medical student at St Thomas' Hospital, London, he did a minimal amount of work. Aided by monthly checks from his father, he preferred to spend his time in art galleries, at the theater, and reading H. G. Wells and George Bernard Shaw.

Later, however, at the London School of Hygiene and Tropical Medicine, he happened to pick up a copy of *Infection and Resistance* by the American bacteriologist Hans Zinsser (4). This fired his enthusiasm for science, a pursuit that contrasted sharply with the helpless, hopeless pill giving he had seen at St Thomas' and which he later called "hogwash."

It was in London, too, that he met Oscar Teague from Harvard Medical School, who arranged an offer of a post there, which Theiler took up in 1922. Once at Harvard, Theiler became friendly with Zinsser and, this being in the midst of Prohibition, was soon exchanging recipes for home brewing.

Thus to yellow fever. Theiler found himself in the midst of a passionate debate about the cause of the condition. Two decades earlier, Walter Reed and James Carroll, working for a U.S. Army Commission studying the disease in Havana, Cuba, had produced evidence that a filterable virus was responsible, but this had failed to resolve the issue to everyone's satisfaction.

In particular, Hideyo Noguchi, working at the Rockefeller Institute, insisted that the culprit was a spirochete. Andrew Sellards, Theiler's boss, thought it was a bacterium, though not Noguchi's spirochete. Theiler, taking a stand unlikely to endear him to his new chief, argued that a virus must be responsible.

He was right. Soon, he had taken a major step forward by making the suspect virus grow for the first time in a laboratory animal by injecting it into the brains of mice. Surviving an attack of the disease himself, he wrote up his work for *Science*, provoking hostile criticism from microbiologists at Harvard and elsewhere. He showed that the virus became attenuated for monkeys when grown in mice and demonstrated that immune serum neutralized the organism. Then, he was tempted away from an unimpressed Harvard by a job at twice his current salary at the Rockefeller Foundation.

Wilbur Sawyer, the tempter, acted with great insight, because Theiler's work had poised him for two major initiatives. The neutralization test made it possible, through the screening of sera, to mount a worldwide survey of the disease's distribution. This was put in hand at once.

Secondly, it seemed likely that an attenuated strain of the virus could be developed as a vaccine. Three years and many thousands of tissue cultures later, a flask labeled 17D yielded the virus that was to form the famous vaccine.

By 1940, field tests were complete, and over the next 7 years, the Rockefeller Foundation manufactured over 28 million doses. A few years later, according to Greer Williams (3), the commuter who had been known to fellow travellers from Hastings-on-Hudson as "the man who lives next door to [baseball player] Alvin Dark" became "the Nobel Prize winner who lives next door to Alvin Dark."

All of this contrasts starkly with the tragic story of Hideyo Noguchi. Born of poor, illiterate parents in northern Japan, he was, at the age of 23, driven by his ambition to seek out Simon Flexner at the University of Pennsylvania. Disappointed not to be offered a job, he nevertheless found support in Philadelphia for work on snake venoms. He also made an unusually deep impression on colleagues at that time.

"He was sensitive, naive, generous to a fault, save where honors were concerned, a spendthrift in time, money and energy, a man of extraordinary drive and industry," writes Paul F. Clark in *Pioneer Microbiologists of America* (1). "We appreciated his childlike simplicity, directness, and the fire-ball intensity of his purpose, and forgave his foibles and weaknesses."

Before Noguchi became interested in yellow fever, he conducted important studies in several disparate fields. He provided the first detailed description of hemolysis triggered by snake venoms and of the damage caused to the endothelium of blood vessels, leading to hemorrhage and edema. His work presaged the development in goats of an antidote to rattlesnake venom, and his meticulous studies resolved uncertainties about the role of *Bartonella bacilliformis* in causing both Oroya fever and verruga peruana.

However, it was yellow fever that proved to be Noguchi's downfall. The problem arose from his absorption, some might say obsession, with spirochetes, which he sought in a wide variety of different infections and tissues. Working at the same time as Theiler, he isolated one such organism that he believed to be the cause of yellow fever and named it *Leptospira icteroides*. However, Theiler and Sellards showed that it was in fact indistinguishable from *Leptospira icterohemorrhagiae*, the agent of Weil's disease (spirochetal jaundice).

Still, Noguchi stuck to his guns. He continued to do so after the Rockefeller Foundation sent a team to the tropics and failed to find his spirochete in yellow fever patients. He even reaffirmed his view in 1927 after Adrian Stokes and colleagues, working in the Gold Coast (now Ghana), transmitted the disease to rhesus monkeys by using material passed through a bacterial filter.

Following the earlier report by Reed and Carroll, this was the definitive confirmation that the actual causative agent was a virus. Soon afterwards (even before the work was published), Stokes contracted the disease and died.

The following year, Noguchi left New York for Accra in the Gold Coast. By now very depressed, he announced, "I will win down there or die." For some months he searched, unsuccessfully, for his spirochete in yellow fever victims. The following May, he too died of the disease.

There is an odd tailpiece to this saga. In 1996, the government of Ghana issued a set of postage stamps to mark the 120th anniversary of Noguchi's birth. This prompted Torsten Wiesel, president of Rockefeller University, to write, "I appreciate the beautiful stamps celebrating the 120th anniversary celebration of the birth of Dr. H. Noguchi issued by the Ghana government. Perhaps you could inform me as to why the Ghana government decided to commemorate Dr. Noguchi."

As reported by S. S. Koide of the Population Council in New York (2), the decision had been taken "because of Noguchi's undaunted devotion and endeavour in identifying the causative agent of yellow fever and in developing a vaccine against this disease in Accra during 1927–8. His presence there had a profound impact on the livelihood of the people in and about the Gold Coast. He exuded hope that inspired confidence that this scourge could be exterminated by executing a frontal attack on this microscopic enemy in a dramatic, 'blood and guts' charge."

One can only speculate as to whether Noguchi would have been pleased to receive this enthusiastic, though largely unwarranted tribute. He might be happier to know that even today his name is remembered in the binomial attached to one of his beloved spirochetes, *Leptospira noguchii*.

References

1. **Clark, P. F.** 1961. *Pioneer Microbiologists of America.* University of Wisconsin Press, Madison.

2. **Koide, S. S.** 2000. Hideyo Noguchi's last stand: the Yellow Fever Commission in Accra, Africa (1927–8). *J. Med. Biogr.* **8:**97–101.

3. **Williams, G.** 1960. *Virus Hunters.* Hutchinson, London, United Kingdom.

4. **Zinsser, H.** 1914. *Infection and Resistance.* Macmillan, New York, NY.

46 René Dubos's
Mirage of Health

Alfred Torrance's *Tracking down the Enemies of Man* (7), Charles-Edward Amory Winslow's *The Conquest of Epidemic Disease* (9), and Peter Baldry's *The Battle Against Bacteria* (2) are just three of many similar titles on my bookshelves. Each tells of past campaigns in our quest to destroy and, if possible, exterminate pathogenic microorganisms. The language is military, the determination absolute.

Many medical microbiologists in the front lines still see their endeavors, quite understandably, in those same robust terms, yet a perusal of the contemporary literature also betokens a significant shift in the philosophy underlying our approach to communicable disease. After decades of successful antimicrobial warfare, developments such as the waning power of antibiotics, the appearance of new infections, and the resurgence of old ones seem to be encouraging a new paradigm.

If this paradigm can be traced to the influence of one man, it is that of René Dubos, who died in 1982 after spending virtually his entire career at New York's Rockefeller Institute for Medical Research (now Rockefeller University). In his book *Mirage of Health* (3), Dubos argued that infectious disease should be seen from an ecological perspective and not simply as the result of collisions between potent agents and susceptible hosts, nor should we strive, through antibiosis and antisepsis, to attain a germ-free existence.

It was in this sense that Dubos used the haunting expression that he selected as the title of his book. He did not argue that health was an illusory concept. He did, however, believe (despite his own role in developing gramicidin, the first clinically useful antibiotic) that deploying ever more potent magic bullets was neither the only nor the most effective long-term strategy for dealing with pathogenic microorganisms. The real keys came from ecology, human behavior, and recognition that microbial and human populations are parts of the same evolving biosphere.

Dubos would be pleased to know that one of the few benefits of our current uneasy relationship with pathogens is the reemergence of familiar yet neglected principles of infection control. Thus, a report in *Epidemiology and Infection* (1) has shown how a potentially serious epidemic was prevented largely by the one simple measure of isolation. The incident was a rubella outbreak that involved four male British soldiers in Bosnia-Herzegovina but that could easily have spread to hundreds of other United Kingdom troops, including nonimmune women; to troops in other nations' peacekeeping forces; and to the local population. All of this was prevented by prompt and rigorous surveillance, isolation, and health education.

Another lesson of this sort was vividly underlined by a report in the *New England Journal of Medicine* in 1994 (5) of an epidemic of pneumococcal disease in a jail in Houston, TX. This was not caused by an exotic, hitherto-unknown organism or a multiply resistant superbug. The crucial ingredients were severe overcrowding and inadequate ventilation, which were responsible for the highest attack rates in certain sections of the prison.

Our forebears of a century ago would not have been at all surprised by this discovery. They were robust believers in the virtues of fresh air. The social reformers who attacked slum housing, and the benefactors who built the isolation hospitals and breezy tuberculosis sanatoria, knew in their bones how to curb the transmission of infection, even before infection was properly understood.

Assessed dispassionately from this broad perspective, present-day Western lifestyles have some extraordinary features. Think, for example, of the cyclic shift in human activity, which provides pathogens with the greatest opportunities for dissemination at the very time of the year when we are most vulnerable.

Commuters into cities press together in the highest densities, with closed windows and excessive heating, during the winter months, when they are most vulnerable to infection. In the summertime, when our resistance is higher, we open our windows and ventilators and enjoy much more space per person. Bus and train carriages are less crowded because people have gone on holiday, many of them seeking isolation on the beach and escape from the crush of the workaday world.

It is odd, too, that our attitudes toward infecting each other are in many circumstances the very opposite of what logic would suggest. We admire tough-minded friends and colleagues who struggle into the office when febrile with a midwinter cold. We make snide remarks about those who stay in bed, and we look upon as socially peculiar anyone who declines to shake hands or kiss at a party because they want to avoid passing on a virus. Perhaps we should be overturning these attitudes and feeling a moral obligation not to infect others, as John Harris and Soren Holm have suggested (4). Social adjustments would be useful, too, such as compensation for lost income.

Another paradox is that we take the greatest risks with food-borne disease in circumstances that pose the highest degree of risk. Whether it is a beach barbecue or a buffet lunch by the hotel pool, we tend to disregard warning signs when we are abroad rather than at home, or at least in our own country. However, this is precisely the time when we are most likely to be exposed to unfamiliar serotypes, pathogens which, because our immune system is not primed to deal with them, are more likely to cause gastroenteritis.

On one occasion, I observed microbiologists at an international conference dinner dining from a vast display of meats, which they all knew had been laid out for several hours beforehand. There was no refrigeration, and the diversity of densely packed dishes offered excellent opportunities for cross-contamination. The only adverse comment came from one participant who pointed out that meats and sauces at room temperature offered enteric organisms even better nourishment than they normally received in the laboratory incubator. Not wishing to offend our hosts, however, he ate heartily with the rest of us.

It may seem perverse to select commuting and conference going as examples of the need for a thorough reexamination of our relationships with potentially pathogenic microorganisms, yet these vignettes could be multiplied many times over, showing that the control of infectious diseases has been and is integrated into our thinking far less profoundly than we might imagine.

When we do adopt the type of perspective which René Dubos first encouraged, cherished ideas come under threat. Are the world's great cities and megacities, with their centers of commercial and cultural excellence, still to be considered among the pinnacles of human attainment? As Richard Horton has argued in *The Lancet* (6), those attributes have increasingly to be considered alongside evidence that the city and its transportation systems is also "a dynamo driving infection." Even the large municipal water system, as Mary Wilson has pointed out (8), has made it possible to infect half a million people with *Cryptosporidium* within a few days.

In a passage in *Mirage of Health* (3), which some today may dismiss as simplistic but most will see as food for thought, Dubos suggested that effective steps in the prevention of disease in future might be motivated by "an emotional revolt against some of the inadequacies of the modern world, and will result from the search for a formula of life more akin to the natural propensities of man."

However, this attitude did not mean a retreat from science. "Far from it," he wrote. "The crusade for pure air, pure water, pure food was at best a naive and often ineffectual approach to the problems of health of the nineteenth century, but it paved the way for the scientific analysis of the factors responsible for the epidemic climate of the Industrial Revolution. Similarly, scientific medicine will certainly define the factors in the physical environment and the types of behavior which constitute threats to health in modern society."

References

1. **Adams, M. S., A. M. Croft, D. A. Winfield, and P. R. Richards.** 1997. An outbreak of rubella in British troops in Bosnia. *Epidemiol. Infect.* **118:**253–257.

2. **Baldry, P.** 1976. *The Battle against Bacteria.* Cambridge University Press, Cambridge, United Kingdom.

3. **Dubos, R.** 1960. *Mirage of Health.* Allen & Unwin, London, United Kingdom.

4. **Harris, J., and S. Holm.** 1995. Is there a moral obligation not to infect others? *Br. Med. J.* **311:**1215–1217.

5. **Hoge, C. W., M. R. Reichler, E. A. Dominguez, J. C. Bremer, T. D. Mastro, K. A. Hendricks, D. M. Musher, J. A. Elliott, R. R. Facklam, and R. F. Breiman.** 1994. An epidemic of pneumococcal disease in an overcrowded, inadequately ventilated jail. *N. Engl. J. Med.* **331:**643–648.

6. **Horton, R.** 1996. The infected metropolis. *Lancet* **347:**134–135.

7. **Torrance, A.** 1929. *Tracking Down the Enemies of Man.* Alfred Knopf, New York, NY.

8. **Wilson, M. E.** 1995. Infectious diseases: an ecological perspective. *Br. Med. J.* **311:**1681–1684.

9. **Winslow, C.-E. A.** 1943. *The Conquest of Epidemic Disease.* Princeton University Press, Princeton, NJ.

47 Ferdinand Cohn, Neglected Visionary

When I read about threats to the species concept posed by modern knowledge of microbial genomes and horizontal gene transfer, I think of Ferdinand Julius Cohn. Though he died over a century ago, the German botanist and bacteriologist would have had no difficulties in coping with pragmatic changes to the very notions that he did much to evolve. He would, I suspect, be wryly amused by today's spirited debates over the blurring of species boundaries.

The real puzzle surrounding Cohn, who was born in Breslau, Silesia (now Wrocław, Poland), in 1828, is why his name is almost totally eclipsed in accounts of the emergence of microbiology by those of Robert Koch, Paul Ehrlich, and Louis Pasteur. He does not, for example, appear in *Microbe Hunters* (1), in which Paul de Kruif paints vivid pen portraits of all the other pioneers. Just as perplexing, *A History of Medical Bacteriology and Immunology*, an apparently scholarly book by W. D. Foster (2), describes Cohn as "one of the founders of bacteriology" but then says nothing whatever about him.

The main sources of this neglect are probably Cohn's personality and his chosen field. Unlike Pasteur, who loved theatricality (exemplified by his public demonstration of the efficacy of chicken cholera vaccine at Pouilly-le-Fort, France, in 1881), Cohn was not a showman. Also, he worked in a discipline which many biologists and historians seem to find rather boring. He was a systematist, classifying organisms into tidy groups.

Ferdinand Cohn's principal achievement, recorded in *Untersuchungen über Bacterien* in 1872, was to show that bacteria could be categorized, like plants and animals, into genera and species. He demonstrated that bacilli, for example, did not transmute capriciously into cocci, nor vice versa.

The experiments he conducted to establish these facts occupied many years of intensive research at Breslau, where he was professor of botany. They went far to disentangle a mass of previous, confusing reports. Far from being of mere academic interest, his findings were vital in firmly establishing the concept of specific etiology: particular infections are caused by particular microorganisms, which do not vary between one case and another but which can be isolated and studied in the laboratory.

We now know, of course, that bacteria and other animalcules do change their forms. Mutation and what the *New England Journal of Medicine* has called "genetic gymnastics" alter their characteristics, providing the raw material for evolution. Still, despite the increasing use of molecular technologies for detecting and identifying pathogens, species labels remain as valid and useful as they were in Cohn's time. Indeed, we still use the basic classification he developed.

It is difficult now to appreciate the visionary importance of Cohn's work for his own time. Hitherto, bacteriologists had meddled with mixed cultures and put forward innumerable theories to explain the alleged pleomorphism of the microbial world. Ernst Hallier, another German botanist and bacteriologist who made major contributions to the subject, was certain that the cowpox (vaccinia) virus (also used for smallpox vaccination) sometimes turned into a fungus. He also claimed that the micrococci responsible for enteric fever were a stage in the life history of the mold *Rhizopus nigricans*.

Cohn swept away such notions and the sloppy experimental work that often went with them. This was an essential step in efforts to relate specific diseases to specific organisms. Cohn was also surprisingly modern in his approach to the classification of bacteria.

Here, too, many writers have failed to appreciate his true significance. When Cohn does receive a footnote on an early page of a microbiology primer, he is usually represented as the man who grouped bacteria simply by their morphology. In fact, he did not apply this criterion rigidly or simplistically. He went out of his way to emphasize that bacteria that look alike may differ from each other in their physiology. He argued, therefore, that metabolism should be an important factor in classification.

It is these insights that persuade me that Ferdinand Cohn would have enthusiastically taken on board successive advances that have since modified our approach to systematics. Consider first the difficulties created for microbiologists by slavish adherence to Ernst Mayr's definition of species as "groups of actually or potentially interbreeding natural populations, which are reproductively isolated from other such groups" (3). I cannot imagine Cohn following those who, because this definition is strictly applicable only to sexually reproducing organisms, have reached the reductio ad absurdum that species do not exist.

Other problems have stemmed from genome mapping over the past decade, which has shown that horizontal gene transfer (between not only strains, but also species) is far commoner than previously supposed. About a quarter of the *Escherichia coli* genome, for example, seems to have been acquired from other species. Again, this need not pose insuperable problems for systematists trying to fathom the relationships between different organisms, nor does it compromise the identification of clinical isolates in hospital laboratories.

Just as Ferdinand Cohn, ahead of his time, grasped that morphology should be combined with physiology in classifying organisms, so he would have welcomed the new computational methods that are beginning to resolve the taxonomic dilemmas posed by gene swapping. He would not, however, be impressed by recent proposals to reject Linnaean binomials altogether in favor of so-called "PhyloCodes."

Practical motives, seen behind Cohn's systematic work, can be discerned in many of his other contributions to microbiology. Together with Pasteur, for example, Cohn denounced the idea of spontaneous generation, but he went further, becoming the first person to demonstrate that organisms such as *Bacillus subtilis* can form spores that are resistant to heat and other physical agents.

He showed that many bacteria can be killed by being boiled but that spores are more resistant than vegetative forms. This was a crucial discovery in invalidating the work of Henry Bastian, primarily an English neurologist but principally remembered for his vigorous opposition to Pasteur. Bastian's apparent demonstrations of heterogenesis were actually attributable to the presence of sporing organisms in infusions supposedly sterilized by heat.

By all accounts, Cohn was also a nice man. It was he who encouraged Robert Koch, at the age of 33, to pursue his historic work on anthrax. Cohn even published Koch's first anthrax paper, in the journal he founded in 1876. Their friendship began during the spring of that year, when Koch, then working in a primitive laboratory at home in Wollstein (a small town in Polish Prussia, where he was a district medical officer), felt that he had solved the major problems of the etiology of anthrax but needed advice on what to do next.

"Esteemed Professor," he wrote to Cohn, "I would be most grateful if you, as the leading authority on bacteria, would give me your criticism of my work before I submit it for publication." Cohn received many such communications from dilettantes and felt pessimistic about this approach by an unknown doctor from an obscure address. Nevertheless, he acceded to Koch's request to visit him in Breslau to demonstrate his experiments.

"Within the very first hour I recognized that he was an unsurpassed master of scientific research," wrote Cohn of his reaction on that first day. Cohn invited other observers to witness subsequent demonstrations and afterwards sent Koch home in a highly elated state. Koch's report of his experiments was complete 3 weeks later, and Cohn published it the following year. However, he did not leave matters there. He continued to help Koch, assisting him to move out of medical practice and establishing him on the scientific career for which he is now renowned.

Ferdinand Cohn deserves our admiration and gratitude for many reasons. When will someone write a full-scale biography of this innovative and prescient man?

References

1. **de Kruif, P.** 1926. *Microbe Hunters.* Harcourt, Brace, and Co., New York, NY.

2. **Foster, W. D.** 1970. *A History of Medical Bacteriology and Immunology.* Heinemann, London, United Kingdom.

3. **Mayr, E.** 1942. *Systematics and the Origin of Species.* Cambridge University Press, Cambridge, United Kingdom.

48 Johannes Fibiger, a Dane to Remember

The year 2013 will see the 100th anniversary of a historic publication by Johannes Fibiger, who, according to traditional ridicule, received a Nobel Prize for the clearly erroneous notion that nematodes cause malignant tumors. Although he worked with rats, the Danish physician believed that parasitology held the key to secrets of the etiologies of some human cancers, too.

These ideas have seemed rather silly to several writers, who have been able to reflect with hindsight upon the false trails and frustrating mirages that have always accompanied cancer research. William S. Beck, in his otherwise excellent book *Modern Science and the Nature of Life* (1), records the Dane's contribution in these words as part of a passage outlining wrong-headed speculations. "Many stories of this kind could be told. In 1926, the Nobel Prize was awarded to a man named Fibiger for 'proving' that cancer was caused by certain small worms." Beck then leaves his readers to chuckle at the man's naiveté.

Now, it is true that by honoring Johannes Andreas Grib Fibiger "for his discovery of the *Spiroptera* carcinoma" the wise persons at the Karolinska Institute (who select the winners in physiology or medicine) implied that he had identified the specific agent of a form of cancer. Perhaps as a consequence of their wording, the authorities became unduly cautious, not only about cancer etiology, but more widely about the risk of prematurity in bestowing their uniquely prestigious accolades. This was probably why Peyton Rous, who in 1911 had reported his discovery of a virus causing cancer in chickens, had to wait 55 years, until he was 87, to receive the 1966 prize for physiology or medicine.

However, Fibiger did not generalize his conclusions to suggest that he had located *the* cause of cancer. Only later commentators have done that. More important, the meticulous Dane has rarely been credited for the considerable impact that his efforts had on the course of experimental oncology.

To appreciate Fibiger's influence, we should remember that in 1907, when he began his cancer studies, the field was in a somewhat confused condition. The English surgeon Sir Percival Pott had deduced that chimney sweeps often succumbed to scrotal cancer because they were exposed continually to coal tars in soot, but repeated attempts in both Europe and the United States to create tumors by rubbing tar into animals' skin had failed. Even after months of application, the skin of rabbits, rats, mice, and other recipients remained normal.

Thus, despite Pott's persuasive work, the experimental study of cancer languished. Rival explanations for the origin of malignancy proliferated with corresponding vigor. The theory that prolonged irritation was a contributory factor emerged strongly from observations of occupational cancers, but far from there being any proof, researchers were totally unable to reproduce malignant tumors, by this or any other means, in the laboratory.

Enter Johannes Fibiger, who followed in his father's footsteps by qualifying in medicine, subsequently worked with both Robert Koch and Emil von Behring, and later held the chair of pathological anatomy at Copenhagen University. He was a founder and editor of *Acta Pathologica et Microbiologica Scandinavica*.

It was an accidental observation that triggered his researches into the causation of malignancy. He noticed, in the stomachs of some rats, tumors which in turn contained a parasitic nematode, later called *Spiroptera neoplastica*. Neither the cancer nor the worm was known previously. Clarifying the relationship between the two demanded a long, painstaking investigation in which Fibiger's scrupulous approach belies the sarcasm of his later detractors.

First, he tried but failed to elicit tumors by feeding rats with either the nematodes or their eggs. Then, he looked into *Spiroptera*'s life cycle and realized that it passed part of its time in the cockroach. The eggs produce larvae in the intestines of this intermediary host, and these then enter its striated muscles, where they become encysted. Rats are not infected by consuming the parasite directly, only by eating infected cockroaches.

Sexually mature worms develop in the rat stomach, in the forepart of which tumors may then appear. These are malignant, sometimes yielding metastases. They are also capable of being transplanted into healthy rats. By revealing this sequence of events in 1913, Johannes Fibiger showed why the tumors he had discovered were so rare: they appeared only when rats ingested the parasite as larvae, and even then not with predictable certainty.

Fibiger's work had two consequences for cancer research. First, it established the experimental study of malignant disease by showing for the first time that cancer could be induced in laboratory animals. Ninety years later, with cancer research reliant on a range of disciplines from epidemiology to molecular genetics, the significance of this shift may appear less than striking. In fact, it marked a historic thrust in the advance of medical science.

Johannes Fibiger's other contribution was to convince scientists that chronic irritation could indeed trigger the emergence of cancer. It was probably both mechanical and chemical irritation from the *Spiroptera* parasites, rather than any specific oncogenicity, that precipitated the development of the rat tumors. Although the explanation of his work is uncertain to this day, its role as a stimulus to others is not at issue.

One direct result was that, in 1915 and 1916, the Japanese oncologist Katsusaburo Yamagiwa succeeded, where others had failed, in producing skin cancer by rubbing coal tar repeatedly into rabbits' ears. We can trace this advance, together with consequent work by Sir Ernest Kennaway at the Royal Cancer Hospital in London, and the trail of research leading to modern concepts of promoters and inducers of malignancy, to the Dane's dogged efforts.

Thus, the 1926 Nobel Prize for Physiology and Medicine went to Copenhagen. It did so not because anyone was convinced that Fibiger had pinpointed *the* agent of cancer, but because his work had laid the foundations for a major forward movement. According to the official record of their deliberations, the mandarins of the Karolinska Institute adjudged that Yamagiwa's studies, though apparently more spectacular, did not have "the same degree of originality" as those of the Dane. For once, the Nobel authorities displayed rare discernment in selecting the scientist to receive their unique award.

Now that we are aware of the roles of human papillomaviruses in cervical cancers, *Helicobacter pylori* in stomach cancers, and hepatitis B virus in hepatocellular carcinoma, Fibiger's studies of *S. neoplastica* no longer seem outré. Nevertheless, we should applaud the rigor of his work and recall his Nobel Prize as the first official recognition that microorganisms play a role in the etiology of cancer.

It would be reassuring to record that the years of controversy surrounding both the great Dane and the 1926 award have ended, but the signs are not hopeful. In some places, indeed, ridicule seems to have been replaced by embarrassed silence. Several prestigious bibliographical reference works have no entry on Fibiger at all. One example is the *Cambridge Dictionary of Scientists* (2). His sole appearance in this book, which contains biographies of over 1,300 individuals, is in a list of Nobel laureates in "psychology or medicine" (sic). This is not impressive.

References

1. **Beck, W. S.** 1957. *Modern Science and the Nature of Life.* Harcourt, Brace, New York, NY.

2. **Miller, D., I. Miller, J. Miller, and M. Miller (ed.).** 2002. *Cambridge Dictionary of Scientists.* Cambridge University Press, Cambridge, United Kingdom.

49 Frederick Twort,
Codiscoverer of Phages

Nearing 70 but determined not to retire, the discoverer of bacterio-phages, Frederick Twort, applied for eight different scientific posts. His potential employers ranged from the London Water Board to a company developing new treatments for bacterial infections. All rejected the distinguished but testy man whose work on phages was to prove pivotal in the development of molecular biology.

Over half a century later, Twort would have been delighted to know that his bacterial viruses are beginning to find uses ranging from the treatment of disease to the tracing of water pollution. At least two such applications are based on genetic modification and thus stem directly from the work of the "phage school" (Max Delbrück and other pioneer molecular biologists in the 1940s).

The first of these innovations, developed at the Medical University of South Carolina in Charleston, has the potential to become a peculiarly potent weapon to attack pathogens. The second, engineered at Lancaster University in Britain, promises to be a much more discriminating tool for identifying sources of pollution than those used hitherto.

There is an odd disparity about the recent and otherwise excellent biographies of Frederick Twort and Félix d'Hérelle, the French Canadian whose 1917 "discovery" of bacteriophages 2 years after Twort helped to

Animalcules: the Activities, Impacts, and Investigators of Microbes

rescue the Englishman from obscurity. Antony Twort's *In Focus, Out of Step* (5) is a lovely portrait of an eccentric polymath who made violins, was a highly skilled amateur radio constructor, designed a more efficient internal combustion engine, and tried to breed the biggest sweet pea in England for a newspaper competition. Despite being an experienced microbiologist, he threw meat and vegetables each day into a large cooking pot of stew that he kept continually on the hob.

Twort's biography of his father surprisingly neglects the part played by phages in the emergence of molecular biology. It does not even mention the classic experiment in which Alfred Hershey and Martha Chase labeled phage protein and nucleic acid and showed that it was the latter that entered bacteria when they were infected.

On the other hand, William C. Summers' *Félix d'Herelle and the Origins of Molecular Biology* (4) goes too far the other way. By its very title, the book suggests that d'Hérelle played a conceptual role in this great chapter of 20th-century science, rather than simply helping to provide one of its most valuable experimental organisms.

Frederick Twort was a brilliant but eccentric man who for much of his career engaged in splenetic and often unreasonable conflicts with Britain's Medical Research Council. His grievances were perhaps heightened by his relative isolation as superintendent of the Brown Institution in London. Founded on the legacy of a rich Dubliner, the purpose of this unique center was to look after sick animals at little or no cost and to conduct research into animal diseases. In the 70 years between its foundation and World War II, when one of Hitler's bombs brought its work to an end, at least a quarter of a million dogs, cats, horses, and other animals were treated there.

However, as superintendent, Twort never had sufficient money or staff to work as he wished. Although the Medical Research Council helped over many years, Twort's gratitude was eclipsed by the anger he felt when those funds were later reduced following less than satisfactory reports on his research.

Much of this work, not only the discovery of bacteriophages, but also the first cultivation of Johne's bacillus and the discovery of the accessory food factor later known as vitamin K, had been of undoubted practical importance, but there was little enthusiasm for his speculations on viruses as the most primitive forms in evolution and his suggestions that he had cultivated them in the absence of living cells.

Although he made no steps in developing the therapeutic potential of phages, Twort would be pleased today to see that this approach is now firmly on the agenda. This follows many decades when the idea was strangely neglected, though with two notable exceptions during the 1980s.

The first was work conducted by Willie Smith and colleagues at the Houghton Poultry Research Station in Britain. They demonstrated that bacteriophages could control diarrhea caused by enteropathogenic *Escherichia coli* in calves, piglets, and lambs.

The second exception was clinical work in Wrocław, Poland. There, phage therapy proved successful in dealing with severe suppurative wound infections that had failed to respond to any other therapy. Perhaps because these findings were published in peripheral journals, they failed to attract much interest in the West. Recent years, however, have seen a minor explosion in therapeutic-phage studies, reflecting the pressing need to develop alternatives to antibiotics because of the burgeoning problem of resistance.

A major landmark was the paper by Carl Merril and colleagues at the U.S. National Institutes of Health and at Exponential Biotherapies in New York published in the *Proceedings of the National Academy of Sciences* (3). They used bacteriophages selected by serial passage for the capacity to avoid elimination from the body for much longer than wild-type viruses, an important advance in overcoming a previous obstacle to effective phage therapy. Used to treat potentially fatal infections in mice, the phages were highly effective against both enteropathogenic *E. coli* and *Salmonella enterica* serovar Typhimurium.

The American Society for Microbiology's 2000 General Meeting marked further advances. One was made by Paul Gulig and coworkers of the University of Florida, Gainesville. They used a particular phage to combat infection with *Vibrio vulnificus*, which can cause a devastating human infection, in vulnerable mice. This development illustrated another likely advantage of phage therapy, its specificity and thus its freedom from the dangers that can arise when antibiotics knock out the body's natural flora.

Animalcules: the Activities, Impacts, and Investigators of Microbes

That potential was underlined in a presentation at the ASM meeting by David Schofield and a team from the Medical University of South Carolina. They genetically modified a phage to code for a bactericidal protein that would then be produced by the target bacterium, a Trojan horse containing the instructions for mass suicide.

More recently, researchers at the Vienna Biocenter in Vienna, Austria, genetically engineered bacteriophages to promote their bactericidal activity while at the same time minimizing the release of endotoxins (2). They believe that their technique could address safety concerns regarding the use of phages as therapeutic agents.

Recombinant phages may also prove to be exceptionally valuable as biotracers to introduce into suspected sources of water pollution in order to monitor their spread. Richard Smith and colleagues at Lancaster University have created a library in which each M13mp18 phage genome contains a unique identification sequence (1). Restriction site polymorphism and other methods are used to identify the phages in water.

This strategy overcomes the two limitations of the small number of phages currently used for this purpose. These are the difficulty of distinguishing them from organisms present naturally (ammunition for defense lawyers in legal cases) and the impossibility of testing several possible pollution sources simultaneously. The Lancaster technique, field tested at an abandoned oil refinery, appears to be a major advance.

The only query regarding the timely exploitation of this new tool for environmental protection is not technical but social. In a world that has recently seen vigorous though irrational opposition to genetic modification, is the deliberate introduction of recombinant phages into rivers and other water systems likely to be publicly acceptable? The same question might be asked regarding therapeutic applications of modified phages.

There is little doubt what Frederick Twort's answer would have been. A believer in "science and efficiency," he demanded that research be unfettered and directed solely by scientists and insisted that "experts in each branch must teach, advise, control and appoint."

Even in the mid-20th century, such sentiments appeared somewhat reactionary. Have we now gone too far in the opposite direction?

References

1. **Daniell, T. J., M. L. Davy, and R. J. Smith.** 2000. Development of a genetically modified bacteriophage for use in tracing sources of pollution. *J. Appl. Microbiol.* **88:**860–869.

2. **Hagens, S., and U. Bläsi.** 2003. Genetically modified filamentous phage as bactericidal agents: a pilot study. *Lett. Appl. Microbiol.* **37:**318–323.

3. **Merril, C. R., B. Biswas, R. Carlton, N. C. Jensen, G. J. Creed, S. Zullo, and S. Adhya.** 1996. Long-circulating bacteriophage as antibacterial agents. *Proc. Natl. Acad. Sci. USA* **93:**3188–3192.

4. **Summers, W. C.** 1999. *Félix d'Herelle and the Origins of Molecular Biology.* Yale University Press, New Haven, CT.

5. **Twort, A.** 1993. *In Focus, Out of Step.* Alan Sutton, Stroud, United Kingdom.

50 Alick Isaacs and Interferon

Just over 50 years ago, a paper in the prestigious but then rather inconspicuous *Proceedings of the Royal Society B* (2) initiated one of the most remarkable stories in modern medical science. The months following Alick Isaacs and Jean Lindenmann's description of interferon, a potentially invaluable antiviral agent, saw a mood of excitement and euphoria. However, these emotions were soon alternating with gloomy skepticism, creating a roller coaster of optimism and pessimism that continued for many years. Only in more recent times has interferon found its rightful place in the control of serious virus infections.

It is an appropriate time, therefore, to salute the achievement of Alick Isaacs, a man of great warmth and tenacity, but also a tragic figure who died at the age of 45 just 10 years after his historic work. He would be as pleased to learn about interferon's hard-won successes as he would be amused to know of the gargantuan muddle that later surrounded the nomenclature of the proteins carrying the name he devised.

Born and trained in Glasgow, Isaacs settled at London's National Institute for Medical Research in 1949 after a period under Macfarlane Burnet at the Walter and Eliza Hall Institute in Melbourne, Australia. Isaacs deeply impressed Burnet both as a man and as a scientist absorbed by a "first rate problem." That problem, which became his life's work in London, was to understand why animal cells infected with one virus become refractory to invasion by another.

In the crucial experiment, conducted with the visiting Swiss researcher Jean Lindenmann, Isaacs added influenza virus, inactivated by heat, to pieces of chorioallantoic membrane from a developing hen's egg. They then recovered fluid from the membrane, incubated it with fresh membrane, introduced live virus, and incubated it again. The virus failed to grow, indicating that something inhibitory, which proved to be a protein, had been transferred in the fluid. Because it explained the well-known phenomenon of interference, Isaacs named the protein interferon (IF, now abbreviated IFN). He and Lindenmann soon realized that it was one of the most powerful biological agents ever known. Clearly, their natural antiviral substance had potentially enormous value as a drug.

That was how both scientists and the media reacted to the news. Once it became clear that the London duo had discovered the human body's first line of defense against many different diseases, a potent antiviral protein that appeared within hours of infection, other researchers began to flock to the field. Popular interest burgeoned, too, and Isaacs was delighted when a Flash Gordon cartoon highlighted his work. "FOUR DEAD, THE OTHERS IN A COMA AND NOTHING WE CAN DO BUT WATCH THEM GO . . . THIS COULD BE IT! INTERFERON! . . . HURRY! . . . WELL DOCTOR? . . . THE INTERFERON WORKS! THE FEVER IS GOING DOWN!"

But three snags soon dampened the euphoria. Since interferon was species specific, the chick product did not work in patients. Human interferon proved much more difficult to make. It appeared in vanishingly small quantities, anyway, and was very difficult to purify. Another 20 years were to pass before an effective method of purification became available.

These difficulties, combined with the publication of papers reporting results due not to IFN but to impurities, created a climate of confusion and uncertainty. Daunted by the technical problems, some scientists cut back on their work, while others lost interest altogether. Some questioned the very existence of the antiviral protein, which they dismissed as "misinterpretation."

Alick Isaacs did not lose faith. He continued doggedly to tackle the tasks of characterizing and purifying interferon, believing it would eventually find therapeutic applications. Then, in 1964, he suffered a stroke, which impaired his speech and left him suffering from acute manic depression. Sadly, that cerebral episode also helped to deepen a growing cynicism about IFN. By the time Isaacs died in 1967, the explosion of research was over. Few virologists even believed in interferon any more.

Isaacs's widow, Susannah, later wrote of the effect of her husband's illness on the climate of thinking about his discovery. "People are very frightened by mental illness, and frightened by mental illness in their own profession. And that made a lot of people very wary. I think some people thought it wasn't even true—that it had just been a wild idea, like the kind of wild idea that people get when they have a manic illness."

Today, interferons are of course lifesavers, widely used to treat conditions such as hepatitis B, hepatitis C, hairy cell leukemia, and chronic granulomatous disease. This transformation owes much to the persistence of a mere handful of scientists who remained convinced of the potential of Isaac's work and helped to solve the considerable technical problems involved in making interferons. They include Derek Burke, Norman Finter, and Barrie Jones in the United Kingdom; Kari Cantell in Finland; Tom Merigan in the United States; and the American researcher Ion Gresser, who did much of his work in France.

Technical barriers aside, the pioneers did not have an easy time in steering a prudent course between the positive, confident endorsement of interferon research and the need for caution. Time and time again, successes with IFN were intermingled with disappointments, as in its failure as an effective treatment or prophylactic for the common cold.

Media coverage was often unhelpful, too. One particularly regrettable episode occurred in 1980, when two young cancer patients in Glasgow were treated with IFN and responded initially, but then died when the supply ran out. The physician who had encouraged publicity for his work was criticized just as heavily as the press by other members of the medical profession.

Fortunately, interferon is now both widely available and successful in clinical practice, with a permanent place in the pharmacopoeia. It is ironic, therefore, that in 1986 one of the IF pioneers, Norman Finter, had to write to *The Lancet* (1) with a warning about the extraordinary confusion that came to surround the most widely used subset of the proteins he helped to develop. Alick Isaacs would indeed have been amused.

Interferons are divided into type I (which includes alpha interferon [IFN-α], IFN-β, and IFN-ω) and type II (IFN-γ). There are 13 different versions of IFN-α. They have names from IFN-α1 up to IFN-α21, with gaps because of past mistakes and artifacts. These proteins can also be made in three different ways. They can be produced by stimulating leukocytes with a virus or other inducer. They can be made in lymphoblastoid cells, the technique developed by Finter and Karl Fantes for Burroughs-Wellcome in the 1970s. Nowadays, they also come from genetically engineered *E. coli*.

This is what led to various varieties of confusion. For example, preparations of ostensibly the same IFN-α contain different proportions of subtypes depending on the particular leukocytes or lymphoblastoid cells from which they were obtained and how they were purified. Moreover, three different recombinant versions of IFN-α2, differing in just 1 amino acid, have been in clinical use.

However, this is not all. Writing in *The Lancet* (1), Norman Finter warned that an "august nomenclature committee" had given additional labels to some interferons. In some cases, this led to the use of at least three alternative names. Thus, Schering-Plough's Intron A was also called IFN-α2 and recombinant IFN-α2b. Glaxo-Wellcome's Wellferon was known as human lymphoblastoid interferon and IFN-αN1, too.

Another consequence was a spate of published errors. Finter gave two examples of results achieved with one preparation that were wrongly attributed to another. Such mistakes would not matter if all types of IFN-α behaved identically in clinical use, but there was increasing evidence to suggest that this was not so, especially in antigenicity.

Adopting George Orwell's dictum, Finter suggested that we should consider all preparations of a particular interferon equal, but some were more equal than others. Clinicians, to be absolutely ambiguous, should specify not only the interferon they believe they are using, but also the cells in which it was produced and the name of the manufacturer. Things seemed a lot simpler when IFN was simply IF.

As to the future, the recent arrival of albinterferon alpha-2b (3) could mark the beginning of an era when chronic hepatitis C and, indeed, a broad range of other diseases are treated, not with interferons as such, but with their genetic fusion proteins.

References

1. **Finter, N. B.** 1996. The naming of cats—and alpha-interferons. *Lancet* **348:**348–349.

2. **Isaacs, A.,** and **J. Lindenmann.** 1957. Virus interference. I. The interferon. *Proc. R. Soc. Lond. B* **147:**258–267.

3. **Subramanian, G. M., M. Fiscella, A. Lamousé-Smith, S. Zeuzem, and J. G. McHutchison.** 2007. Albinterferon alpha-2b: a genetic fusion protein for the treatment of chronic hepatitis C. *Nat. Biotechnol.* **25:**1411–1419.

51 Dissenters: Max von Pettenkofer and Friedrich Wolter

Scientists who repudiated Louis Pasteur's views on spontaneous generation, like the physicians who challenged Joseph Lister's innovations in antiseptic surgery, are readily dismissed as rather pathetic footnotes to the history of microbiology. They were too inflexible, too lazy, or simply too stupid to acknowledge the great advances of their day.

No doubt such adjectives are appropriate for some of the awkward characters who, decades ago, questioned developments whose intellectual basis and practical significance we now fully appreciate; but not all. Other dogged questioners were behaving as true scientists, submitting novel concepts to rigorous criticism and demanding more definitive evidence before they abandoned established practices and ideas.

That is why televised and cinematic representations of the great turning points of medical science can be so irritating—and inaccurate. Over the years, I recall portrayals of Ignaz Semmelweis's measures to prevent puerperal fever in Vienna in the 1840s, and of Lister's introduction of the carbolic acid spray in Glasgow in the 1860s, which lampooned all of their opponents as near idiots. Such blanket dismissals are unreasonable.

Consider Max von Pettenkofer and Elie Metchnikoff. Around the turn of the century, in Munich and Paris, respectively, they drank cultures of *Vibrio cholerae* from patients who had died of cholera. The pair did so,

Animalcules: the Activities, Impacts, and Investigators of Microbes

bravely, foolishly, or logically, according to one's perspective, as part of a campaign against the triumphalist view that Robert Koch's discovery of *V. cholerae* provided a complete answer to the etiology of the disease. Both survived the experience. Although enormous numbers of vibrios were recovered from their intestines, neither experimenter developed cholera.

It would be easy to see these two incidents as extremely lucky escapes that retarded progress toward a real understanding of cholera by appearing to invalidate Koch's discovery. Pettenkofer and Metchnikoff certainly felt that self-experimentation had vindicated their views. In the long term, however, as René Dubos observed in his splendid book *Mirage of Health* (1), their work clearly and usefully established that factors such as nutrition can have a profound influence on communicable disease.

The simplistic view, prevalent at the time, that an infection was the inevitable outcome whenever its specific agent encountered a host, was a gross oversimplification. This is a lesson that needs to be relearned in every generation, as exemplified by some of the disagreements during the 1990s about the cause of AIDS.

When, though, does constructive resistance to novel ideas become destructive obstinacy? Certainly, I suggest, in the case of the Hamburg-based hygienist Friedrich Wolter, who could also be termed the last of the anticontagionists. More than 100 years ago, he was spurred by the publication of an analysis of the 1892 Hamburg cholera epidemic, written by Koch and others, to produce his own counteranalysis. Koch's 1896 report clearly reflected the supreme importance of *V. cholerae*, which he had discovered 13 years earlier, in causing the disease, but Wolter was not convinced. He remained unmoved year by year, and indeed decade by decade, thereafter. As many other bacteria, viruses, fungi, and protozoa were incriminated as the specific agents of other diseases, his disbelief in contagion grew correspondingly stronger.

Even as recently as 1944, the year in which Wolter died, he was still publishing critiques of the role of microbes in communicable diseases. Four decades after the momentous work of pioneers such as Robert Koch, Paul Ehrlich, and Ronald Ross had been recognized by the award of their Nobel Prizes, Wolter's heterodox papers continued to appear, and in reputable journals, too.

Friedrich Wolter was one of Pettenkofer's most fervent disciples in believing that climatic conditions, and especially fluctuations in the levels of ground water, were crucially important in the causation of cholera. In 1893, at the age of 30, he sent Pettenkofer an account of cholera epidemics in Hamburg that apparently vindicated the "soil doctrine." The older man was delighted, and it was he who financed publication of Wolter's counterblast to the 1896 report of Robert Koch.

During the second half of the 19th century, when he wrote over 200 papers and 20 monographs, Pettenkofer's "soil doctrine" was by no means as unfashionable as might be imagined today, nor was he wholly opposed to the notion that germs might play a part in disease, as some writers have claimed. What he argued was that the principal causes of cholera and other conditions were miasmata emitted by certain terrains. Though ill-founded, this belief led Pettenkofer to campaign successfully for adequate sewage systems in Munich.

Nevertheless, supporters of the soil doctrine had dwindled to a tiny minority by the time Pettenkofer died by suicide in February 1901. At that point, the difference in critical temper between him and Friedrich Wolter became transparently clear. Though embracing an incorrect thesis, Pettenkofer had both conceded a possible role for microorganisms in disease and encouraged adherents of the germ theory to produce better evidence for their views. Wolter, on the other hand, became increasingly extreme in rejecting the germ theory, especially after Pettenkofer's death.

In 1910, Wolter published a review stating that Koch's and Pettenkofer's theories were in blatant contradiction to each other, denying that the Broad Street pump played any part in the 1854 cholera outbreak in London's Soho, and repudiating John Snow's meticulous analysis of that event. Four years later, a further publication dismissed the role of microorganisms, not only in cholera, but also in dysentery, typhoid fever, and typhus.

Amazingly, the crusade continued into the 1930s and 1940s, long after specific etiology had become one of the central concepts of modern medical science and had triggered many practical triumphs in the conquest of disease. Wolter stuck to his guns, notwithstanding events such as Fred Griffith's work on the transformation of pneumococci, the development of penicillin by Howard Florey and his Oxford colleagues, and the introduction of vaccines against smallpox, diphtheria, tetanus, tuberculosis, and many other conditions.

Among a prolific flow of Wolter's papers, which continued to appear in major journals, were one seeking to demonstrate that poliomyelitis was a "soil disease" and another attributing epidemic hepatitis to a "gaseous, toxic cause of disease developing from an unhealthy soil." In his final paper, he rejected streptococci as the cause of scarlatina, which he attributed instead to variations in ground water, climate, and sunspots.

Writing in the *British Medical Journal* (2), Norman Howard-Jones saw these as the writings of a "mentally deranged author" and expressed legitimate dismay that they were accepted for publication. That, indeed, is the strangest feature of the Wolter affair. It is neither surprising nor scandalous when a composer, producing music in a long-outmoded style, finds a publisher willing to publish it. The tolerance of editors for demonstrably erroneous ideas, and their willingness to disseminate them at the expense of papers describing genuine scientific advances, is a very different matter.

References

1. **Dubos, R.** 1960. *Mirage of Health*. Allen & Unwin, London, United Kingdom.

2. **Howard-Jones, N.** 1980. Friedrich Wolter (1863–?1944): the last anticontagionist. *Br. Med. J.* **180:**372–373.

52 Gerhard Domagk and the Origins of Sulfa

On 11 December 1943, Winston Churchill flew from Cairo, in Egypt, to the North African city of Tunis to spend a few days with Dwight D. Eisenhower at his "White House" near the ruins of Carthage. The British Prime Minister had already completed a complex series of meetings, conferring with the Chinese general Chiang Kai-shek in Cairo and with President Franklin D. Roosevelt and the Soviet leader Josef Stalin in Tehran. Their agenda was an evolving plan for the D-Day landings to regain France from the Germans.

Overweight and a heavy smoker and drinker, Churchill was also overworked and exhausted. His flight to Tunis was delayed, and by the time he reached Eisenhower's villa, he had a sore throat and was feeling unwell. In the coming days, his temperature rose rapidly, his condition deteriorated, and a portable X-ray machine confirmed that he had contracted pneumonia. When Churchill suffered two bouts of atrial fibrillation, his physician decided to give him not only digitalis for the heart problem, but also a new drug to combat the pneumonia. It was one of the sulfonamides, called M&B 693, named after the British drug company May & Baker, which had developed it.

The treatment was successful, and 2 weeks after becoming ill, the British premier was able to fly to Marrakesh and then home to London, where he continued to refine the D-Day plans. While he had been delighted to be allowed a shot of brandy with the M&B 693, there is little doubt that the novel sulfa drug defeated the pneumonia and probably saved his life.

Given the eminence of the patient and the dramatic nature of this and many other, earlier cures, it seems remarkable that there is now so little awareness of the impact made by sulfa drugs on the conquest of infectious maladies. Everyone knows about Louis Pasteur, the first scientist to devise specific vaccines against pathogens, and about Alexander Fleming's work on penicillin, but who recalls the name of Gerhard Domagk, the German biochemist responsible for the inception of the sulfa drugs during the 1930s? Even today's physicians can be hazy about this crucial episode between the historic studies of Pasteur, Robert Koch, and Paul Ehrlich and those of the Oxford researchers who turned Fleming's laboratory observations into the revolutionary advance of penicillin therapy.

It was in an effort to make good this deficiency that Thomas Hager decided to write *The Demon Under the Microscope* (1). He succeeded extraordinarily well. Do not be deceived by the hyperbole and raciness of the cover ("The Nazis discovered it. The Allies won the war with it. . . . This incredible discovery was sulfa"). Hager's book is a well-researched chronicle which goes a long way to justify his claim that the key event in the history of the treatment of bacterial disease was not the emergence of penicillin in the early 1940s but the discovery of sulfonamides a decade earlier. If we accept his plea that all antimicrobials should be described as antibiotics, and not just those made by living organisms, then "sulfa" was the first truly revolutionary antibiotic.

Not least of Hager's skills is that, alongside some admirable science writing, he illuminates the social context of the research. Especially vivid is his portrayal of the appalling conditions which Gerhard Domagk encountered as a youthful medical assistant on the Eastern Front in World War I and which inspired him to pursue his medical studies in the hope of developing effective therapies for conditions such as gas gangrene. "He witnessed more operations in two years than many surgeons see in a lifetime," Hager writes, "helped set compound fractures, the bones bristling through the skin; used magnets to search for pieces of

shrapnel; watched surgeons runs their fingers down the insides of intestines, probing for holes; assisted with countless amputations, threw the
severed arms and legs onto a growing pile in a side room."

Young Domagk's key observations were on the dreadful consequences when even the most heroic and apparently successful surgery
allowed *Clostridium perfringens* to invade the incision site and foster
foul-smelling, potentially fatal wound infections. Gas gangrene was furi-
ously contagious, capable of killing half of the patients in a postoperative
ward within a few weeks, and there were other bacterial enemies, too,
all contributing to the massive fatality rate even among soldiers who
survived the traumas and stresses of battle. A quarter of a century later,
Domagk was to write in his diary: "The real birth of chemotherapy as far
as I am concerned took place in the Great War of 1914–18 when I swore
an allegiance with my fallen comrades. Those were my first principles
and they still are today. They stand over me like a shining star."

Though he worked briefly as a pathologist at the universities of Greifs-
wald and Munster, Domagk discovered the sulfa drugs during a long
career at IG Farbenindustrie, formed from a merger between Bayer and
BASF in 1924. The key discovery, that the sulfonamide-containing Pron-
tosil red could control streptococcal infections in mice, came in 1932.
However, difficulties in replication delayed its publication by about 3
years.

Once the work did become public, and other laboratories and other
countries joined the search, a string of different sulfa drugs appeared,
and the range of curable infections increased apace. By 1940, one sul-
fonamide or another had become standard therapy for pneumococcal
pneumonia, childbed fever, erysipelas, streptococcal infections, and
the commonest forms of meningitis. Soon urinary tract infections, tra-
choma, chancroid, mastoiditis, otitis media, and gonorrhea joined the
list. With all combatants carrying or having rapid access to the new
wonder drugs, death rates from meningococcal meningitis plummeted
in World War II compared with those in World War I.

Two grim ironies may, in different ways, explain why this otherwise
momentous work, which spawned enormous benefits for mankind,
did not receive greater public acclaim. First, the Nazis would not allow
Domagk to accept the Nobel Prize he was awarded in 1939 (though he
did receive just the medal in 1947). Secondly, IG Farben became the

major manufacturer of the Zyklon B poison gas that was used to kill Jews at Auschwitz.

No doubt the advent of penicillin, with its potency and the associated publicity, is the principal reason why the story of Gerhard Domagk and the sulfonamides has been painted out of popular histories of the conquest of pathogenic bacteria. Still, as Thomas Hager showed, the earlier chapter had at least as great an impact on human history in a variety of different ways.

"Just as important as its role in curing any disease, sulfa cured the medical nihilism of the 1920s, dissipating the prevailing attitude that chemicals would never be able to cure most diseases," Hager wrote. "Sulfa proved that magic bullets were possible, encouraged their discovery, established the research methods needed to find them, framed the legal structure under which they would be sold, and created the business model for their development."

Thomas Hager has performed an invaluable service, not only in resurrecting from obscurity a crucial episode in our battles against communicable disease, but also in emphasizing its wider social and industrial significance. What a pity that he mars all of this by misidentifying the three scientists who are famed throughout the world for the discovery and development of penicillin. He tells us that Alexander Fleming (a Scot) was Australian and that Howard Florey (an Australian) and Ernst Chain (a German Jew) were British. Oh, dear.

Reference

1. **Hager, T.** 2006. *The Demon Under the Microscope.* Harmony Books, New York, NY.

53 Cecil Hoare's
Eponymous Organism

Attachment of one's name to a trypanosome isolated from a Brazilian crocodile seems a somewhat ambivalent honor, like being remembered for lewisite, the bowie knife, or brucellosis, but biologists are funny people. Microbiologists, in particular, delight in having their surnames incorporated into the binomial labels of even the most virulent of parasites. When first names, rather than surnames, are used in this way, the implication is of special affection, and the recipient's pride is correspondingly greater.

Thus, Cecil Hoare derived as much quiet satisfaction from an honor bestowed on him in 1977, when he was aged 85, as he did from the many other scientific awards he received during a remarkably long career. Since that year, Hoare's name has been linked with the parasite *Trypanosoma cecili*, isolated from the South American cayman, *Caiman crocodilus crocodius*. The appellation was recorded for the first time by the trypanosome's discoverer, Ralph Lainson of the Instituto Evandro Chaga, Belem, Brazil, in *Protozoology* (4), the entire volume of which formed a Festschrift in recognition of one of the most eminent protozoologists of all time.

Cecil Hoare was born as a British subject in Middelburg, Holland, on 6 March 1892. His first language was therefore Dutch. His second was Russian. Taken to Russia at the age of 6, he was educated at the Imperial

University of St. Petersburg, where he specialized in zoology and graduated in 1917. The revolution soon thwarted a promising career until 1920, when the British government agreed to repatriate him.

In London, he was befriended by Clifford Dobell, author of the delightful *Antony van Leeuwenhoek and His "Little Animals"* (2). With Dobell's help, Hoare secured a grant from Britain's newly established Medical Research Council to work at the Wellcome Bureau of Scientific Research. Thus began an association that continued until well into Hoare's 80s, when research initiated by him proceeded at the Wellcome Research Laboratories, Beckenham, United Kingdom, toward a vaccine for Chagas' disease.

Like Dobell, a gentle man who insisted on keeping his monkeys near him in the laboratory rather than in an animal house so that he could be sure they were looked after well, Hoare was a solitary researcher. He rarely had a technician working for him, and only 13 of his massive bibliography of 179 publications had multiple authors. It seems unlikely that any individual scientist will ever again be responsible for such a formidable succession, and such a variety, of outstanding discoveries.

In these days, when relevance and payoff count for so much in scientific research, the practical value of much of Cecil Hoare's work, which he always insisted was directed largely toward the protozoa rather than the diseases they produce, is worth recalling. Thus, one of his earliest projects concerned the flagellates that live inside another parasite, the sheep ked. Hoare identified the organism concerned, *Trypanosoma melophagium*, and demonstrated its life cycle and transmission by the ked. However, this knowledge was of more than academic interest: 80% of British sheep were naturally infected with trypanosomiasis at that time.

Between 1927 and 1929, Hoare worked in Entebbe, Uganda, where he discovered the life cycle of another crocodile parasite, *Trypanosoma grayi*. The developmental stages of this organism in the tsetse fly were already known, but it was Hoare who demonstrated their relationship to crocodile trypanosomes. During the 1930s, he pioneered important work on pathogenic trypanosomes of pigs, cattle, and monkeys. A particular interest was *Trypanosoma evansi*, the agent of surra, which is purveyed among horses by horseflies. Circumstantial evidence suggested that it originated from another species, *Trypanosoma brucei*, and this led Hoare to propose that camels played a major role in its evolution. Much later in life, he conducted a historical-geographical study tracing the caravan routes from tropical Africa along which he believed surra spread to other parts of the world.

Apart from these and scores of other contributions to our knowledge of trypanosomes, Cecil Hoare originated key lines of research on many other groups of microorganisms. He concluded that *Entamoeba histolytica* was represented by two distinct races, one a cosmopolitan commensal and the other a potential pathogen, usually restricted to hot countries. This hypothesis, now widely accepted, implies that symptomless infections with the organism in temperate regions do not need to be treated.

As Tony Duggan, then director of London's Wellcome Museum of Medical Science, observed in his introduction to the Festschrift, it was difficult to believe that Cecil Hoare retired from his Wellcome post in 1957. It was after that date that he produced his masterwork, *The Trypanosomes of Mammals* (3). This 749-page "monograph," which became a bible for medical and veterinary workers dealing with trypanosomiasis, appeared in 1972. Still, its octogenarian author also found time that year to publish scientific papers in the *Journal of Tropical Medicine and Hygiene* and the *Proceedings of the Royal Society of Tropical Medicine and Hygiene*.

"As a person Cecil Arturovitch (as we sometimes jokingly called him) was the epitome of the quiet, reserved, single-minded, precise scientist with an occasional twinkle in his eye," wrote the malariologist Leonard Bruce-Chwatt in *The Lancet* (1) when Hoare died in 1984. "Though an accomplished linguist, his soft spoken English retained a trace of Russian accent. Some people thought him distant, but his friends knew the extent of his loyalty to them and his unassuming kindness."

Cecil Hoare always retained a keen interest in protozoology in Russia, particularly its historical aspects. He showed, for example, that P. F. Borovsky was the first person to recognize *Leishmania tropica* as a protozoon. It was thus appropriate that his Festschrift contained three papers by Soviet scientists in Russian (with translations). There were also contributions from researchers in Germany, Upper Volta, Brazil, Costa Rica, Switzerland, India, the Gambia, and the United States, as well as Britain—an apt reflection of the breadth of Hoare's work.

However, the most intriguing item in the entire volume is a list of Hoare's 179 original contributions to the literature. Just one of those papers described a new *Grahamia* isolated from gerbils, an *Entamoeba* from the goat, mammalian trypanosomes in caterpillars, a ciliate from the Indian rhinoceros, and two other original laboratory observations on trypanosomes. In modern times, when researchers can be tempted to write up the same discovery three times for three different journals, such disregard for the great paper chase seems barely credible.

As for Cecil Hoare's religious and political outlook, this is most appropriately taken from the autobiography, as yet unpublished, that he prepared in 1975 and revised in 1980:

The study of natural history and sciences, as well as a comprehension of modern cosmological ideas, have resulted in the adoption of a materialistic philosophy and concept of the universe (Weltanschauung) which to my mind is incompatible with any form of religion that demands blind faith: "credo quia absurdum" was therefore unacceptable. Socialism in its ideal, theoretical form had a certain attraction for me but, having witnessed it working in practice, I became disillusioned and ended up becoming politically agnostic. The chaotic situation in the present fin de siecle has forced me to adopt a cynical attitude to current political and social trends. I accept the logic of dialectical materialism, but I reject its corollary, expressed in the dogmatic theory of historical materialism with its political and economic implications.

References

1. **Bruce-Chwatt, L.** 1984. Cecil Hoare. *Lancet* i:533.

2. **Dobell, C.** 1932. *Antony van Leeuwenhoek and His "Little Animals."* Harcourt, Brace, and Co., New York, NY.

3. **Hoare, C. A.** 1972. *The Trypanosomes of Mammals.* Blackwell Scientific Publications, Oxford, United Kingdom.

4. **London School of Hygiene and Tropical Medicine.** 1977. *Protozoology*, vol. 3. *Festschrift in Honour of C. A. Hoare, F.R.S. on the Occasion of His 85th Birthday.* London School of Hygiene and Tropical Medicine, London, United Kingdom.

54 Ants and Fred Hoyle's Challenge to Darwinism

The distinguished English astronomer Fred Hoyle, who died in 2001 at the age of 86, came to believe that terrestrial life did not originate spontaneously in the primordial soup or evolve by mutation and selection. He favored instead an idiosyncratic version of panspermia, according to which living organisms reached Earth from space (perhaps carrying coded nucleic acid messages from their senders) and developed in a non-Darwinian fashion thereafter. He and his collaborator, Chandra Wickramasinghe, believed that interstellar viruses were continuing to arrive here, causing otherwise unaccountable epidemics of diseases such as influenza and cholera.

In Hoyle's view, biologists wedded to neo-Darwinism were victims of a "junkyard mentality." Imagine, he suggested in *The Intelligent Universe* (3), a junkyard containing all the bits and pieces of a Boeing 747, dismembered and in disarray. "A whirlwind happens to blow through the yard. What is the chance that after its passage a fully assembled 747, ready to fly, will be found standing there? So small as to be negligible, even if a tornado were to blow through enough junkyards to fill the whole universe."

Hoyle's analogy is a foolish caricature of the conventional understanding of life's origin and evolution and is arguably best forgotten, except for two things. First, is it not strange that, ever since Louis Pasteur

Animalcules: the Activities, Impacts, and Investigators of Microbes

firmly established that life on Earth did not emerge spontaneously from nonliving substances, scientists have spent a considerable amount of time trying to find out exactly how it did emerge spontaneously from nonliving substances?

Secondly, the difficulty some people face in accepting neo-Darwinism springs not from the intellect but from the imagination. This is evidently a blind spot for biologists seeking to reason with those drawn to the anti-Darwinian musings of Fred Hoyle and others. They fail to recognize that a skeptic may be able to take on board the central conclusions drawn from our huge corpus and variety of evidence for organic evolution while remaining intuitively unconvinced that it really did happen this way.

Perhaps protagonists for neo-Darwinism, despite their certitude, should be more open about their own problems in seeing the "blind" selection of random changes at the heart of evolution. In addition to questions about the emergence of intricate structures, such as the eye, some of the greatest of these difficulties concern behavior—for example, bird migration, foraging bees, and altruism.

Though a convinced Darwinian, I have long found the cultivation of fungus gardens by attine ants (members of the tribe Attini) to be espe-cially challenging. First described by the English geologist Thomas Belt in 1874 in *The Naturalist in Nicaragua* (1), ants' fungus farming has shown ever more complexity as ever more intimate research has been applied to the phenomenon. Later insights, published by Stig Rønhede and colleagues at the University of Copenhagen in Denmark (4), have added further sophistication to the applied microbiology practiced by these dedicated creatures.

Leaf-cutting ants of the genera *Atta* and *Acromyrmex*, which are dominant herbivores and agricultural pests in tropical and subtropical America, have been studied particularly closely. The workers spend much of their day snipping off fragments of fresh leaves, which they carry to their underground fungus gardens and chew into pulp. Although they drink the plant sap, they do not themselves eat the material. Instead, they inoculate it, after adding a little feces, with some of the *Leucoaga-ricus gongylophorus* organisms already growing in established parts of the fungal garden.

As the garden matures, the mycelium produces bundles of swollen hyphal tips known as gongylidia. The worker ants chew these and ingest the cytoplasm to supplement their diet of sap. More importantly, they also feed gongylidia to the next generation of larvae, which grow and develop solely on their fungal food. When a new colony is to be formed, the queen takes along a pellet of fungus to initiate her new garden.

Research over the past quarter century has revealed many astonishing details of this underground industry. They include insights into the elaborate manuring regimes which the ants use to optimize the yield of fungus and their incorporation of *Streptomyces* and other antibiotic producers to combat the proliferation of unwanted molds, "weeds" in the fungal garden.

Recently, particular interest has centered on the question of why attine ants carefully deposit droplets of feces on every snippet of leaf before inoculating them with *L. gongylophorus* and placing them in their garden. Some years ago, feces from *Atta colombica* were found to degrade chitin, pectin, starch, xylan, and carboxymethylcellulose. Moreover, *Atta texana* feces contained proteases identical to those in cultures of the symbiotic fungus.

What Stig Rønhede and collaborators wanted to know was whether the ants' behavior in using fecal droplets when maintaining their fungal gardens had evolved (we must not be teleological, must we?) because incorporation of these enzymes bestowed explicit benefits. If so, had the enzymes evolved so that they were protected from degradation when passing through the ant gut? Alternatively, was the presence of the enzyme activities in the feces insignificant, simply a passive consequence of some proportion of their molecules surviving digestion?

Using isoelectric focusing and specific staining, the Copenhagen team first confirmed that feces from two species of ants, *A. colombica* and *Acromyrmex echinatior*, contained carboxymethylcellulases, laccases, proteases, and pectinases (both esterases and lyases). They then clearly established that the enzymes came from the symbiotic fungus, rather than from the ants themselves.

Their major finding, however, emerged from a comparison of the specific activities of the enzymes in the fungus and in the fecal droplets that the ants deposited on their leaf fragments. These measurements indicated that the enzymes were not only protected, but quite possibly concentrated, too, as they passed through the ants' intestinal tracts.

What appears to have evolved is a pattern of behavior in which ants consume fungal cytoplasm containing degradative enzymes, which appear later in their feces and are thus added to new leaf fragments, where they facilitate the growth of new mycelium. Rønhede and his team speculated that the carboxymethylcellulases and pectins might help to initiate hyphal colonization by attacking plant cell walls. Pectinases, which have a macerating effect on plant tissue, probably accelerate the early stages of degradation.

These two groups of enzymes are likely to be of little or no digestive assistance to the ants themselves because they ingest only plant sap rather than whole cells. This is a further indication that the ants' "purpose" (teleology really is hard to avoid) in adding feces to their leaf fragments is as an aid in the development of their fungal gardens.

Jared Diamond has pointed out (2) that human agriculture, which began around 10,000 years ago in the Fertile Crescent and China, was anticipated by fungus-growing New World ants some 50 million years ago. In more recent times, much has been learned about the coevolution of attine fungi, essentially ancient clones, propagated vegetatively, and their hosts.

We understand little, however, about the evolution of the ants' behavioral patterns, which have led to such a sophisticated and precise system of subterranean agriculture. Though I cannot quite see it, I have no doubt that neo-Darwinian mutation and selection were the driving forces. After all, what is the alternative?

References

1. **Belt, T.** 1874. *The Naturalist in Nicaragua.* John Murray, London, United Kingdom.

2. **Diamond, J.** 1998. Ants, crops, and history. *Science* **281**:1974–1975.

3. **Hoyle, F.** 1983. *The Intelligent Universe.* Michael Joseph, London, United Kingdom.

4. **Rønhede, S., J. J. Boomsma, and S. Rosendahl.** 2004. Fungal enzymes transferred by leaf-cutting ants in their fungus gardens. *Mycol. Res.* **108**:101–106.

55 Pioneers of
American Microbiology

Professor Conn exhibited some cultures of a highly variable Micrococcus which he had isolated many times from milk. Its color ranged all the way from a snow white to a deep orange, and in power of liquefying gelatin it ranged from a form that liquefied with great rapidity to one that had apparently no liquefying power. All these varieties, with numerous intermediate stages, have been found in nature and are not the result of cultivation. Professor Conn showed, however, what a great change can apparently be produced by a simple process of selection. (6)

The abstract of Herbert W. Conn's talk at the first meeting of the Society of American Bacteriologists (SAB) in New Haven, CT, from 27 to 29 December 1899 reflects the simple, active, limpid prose in which much science was then recorded. It also embodies themes that have continued to interest bacteriologists over the past century: the chemical activities of bacteria, differences between their behavior in the laboratory and in the outside world, and their variation and selection.

From the standpoint of medical and much practical bacteriology, the program of that inaugural gathering, held at Yale Medical School, was surprisingly comprehensive. It embraced nomenclature and systematics; sterilization and antiseptics; sewage treatment; dairy bacteriology; the

Animalcules: the Activities, Impacts, and Investigators of Microbes

testing of water, milk, and canned food; investigations of the organisms of plague, tuberculosis, typhoid fever, and actinomycosis; "the so-called fermentation of tobacco"; and even a "new" pathogenic fungus.

Given that Martinus Beijerinck and Serge Winogradsky were then well into their stride in Europe, nonmedical bacteriology seems to have been underrepresented. Nevertheless, the new society's president, William T. Sedgwick, more than compensated for any such imbalance in his fine after-dinner address, which embraced the subject in its broadest sense.

"Bacteriology is a child of the 19th century. It is the offspring of chemistry and biology, enriched by physics with the gift of the achromatic microscope," he began. After cohering into a single vision insights drawn from studies of fermentation, putrefaction, organic decomposition, nitrification, and epidemic disease, he paid handsome tribute to the work of Louis Pasteur.

"Though not exactly a chemist, he was able to meet chemists on their own ground. Though not exactly a microscopist, he was highly trained in physics and mineralogy, and thus quickly became a master of the microscope," Sedgwick said. "Time has proved beyond all peradventure that the foundations laid by Pasteur were laid solidly and securely." Inspired by the Frenchman, "a host of eager and enthusiastic workers" had now thrown themselves "with intense zeal into the study of the micro-organisms which constitute the field of micro-biology."

Throughout the 1899 proceedings, perceptive comments about how bacteria behave and evolve intermingled with frustrating ignorance concerning the underlying mechanisms. "The evolution of parasitic from saprophytic forms is a very slow and gradual process," said Theobald Smith. "Special advantages which a certain environment may offer for frequent passages through susceptible species may give certain saprophytes an impulse towards a parasitic existence."

Smith speculated that some organisms that did not originally cause disease had "fighting characters," such as the capacity to produce toxins, which enabled them to become pathogens. However, "the degree of change that can be impressed upon any bacteria probably depends largely on the specific structure of the organism." Moreover, "processes of conjugation and other sexual phenomena, such as are found among protozoa, are unknown."

Little progress in understanding these matters is apparent from the agenda for the SAB's 25th meeting in 1923, also held at the Yale School of Medicine in New Haven. Nevertheless, the program shows clear evidence that both the Society and the subject itself had expanded vigorously.

Though the gathering again took place from 27 to 29 December, there were 84 talks, more than three times the number in 1899, and the menu not only ranged widely over scientific topics, such as thermophiles, the Twort-d'Hérelle phenomenon (bacteriophages), and the migration of bacteria in an electric field, but also embraced professional challenges.

One hot area was the need for practical standards and replicability, ranging from the accuracy and capacity of the platinum loop to a proposed limit of 10,000,000 bacteria per gram for the condemnation of hamburger steaks.

Speaking of the Committee on Bacteriological Technic, which he chaired, Harold J. Conn (son of Herbert Conn) said that it was "not the function of the Society to standardize technic in the sense of adopting official methods. . . . The activities of this committee should, on the other hand, be concerned with the development of new technic.

"New technic cannot be developed in a logical way, however, without standardization of material and supplies. Accordingly the committee has given assistance in this line, notably in cooperation with other societies in the standardization of biological stains."

However, this was not so simple. In another talk, Conn described a project in which the SAB committee had distributed to several centers slides prepared in one laboratory, together with samples of stains and directions. The results proved to be slightly more constant than in a previous exercise with less stringent controls. However, "there was enough variation to show that certain organisms are really Gram variable; and that it is almost impossible, if not quite so, to standardize the technic sufficiently to make all results agree."

Taxonomy and nomenclature were other pressing concerns. Robert S. Breed and Margaret E. Breed reviewed a catalogue of confusion over the binomial used to describe "the type species of the genus *Serratia*, commonly known as *Bacillus prodigiosus*" and argued strongly for the name *Serratia marcescens*. However, they then relented, suggesting that those who found it hard to give up a familiar epithet could continue to use "prodigiosus as a familiar or trivial name if desired."

Likewise, David H. Bergey highlighted "the inappropriateness of the name *Streptococcus hemolyticus*" when at least eight distinct species had been clearly identified by agglutinin absorption tests. "A strong plea is made for the application in practice of the recent advances in our knowledge of the specific characters of the hemolytic streptococci, especially in the intelligent use of homologous and mixed streptococcus vaccines and of specific monovalent or of polyvalent antiserum containing antibodies against several species of hemolytic streptococci."

The reliability of serological tests for syphilis was the subject of three papers, two of them by Reuben L. Kahn. His flocculation test was then rivaling the Wassermann complement fixation test as a more rapid method of diagnosing syphilis, though both were subsequently superseded.

Then aged 35, just 1 year younger than Kahn, Selman Waksman was another speaker who delivered two papers. The first, given in conjunction with Robert L. Starkey, was on methods of encouraging the growth of various categories of organisms in soil. They reported that the addition of dried blood brought about "an extensive development of fungi, bacteria and actinomycetes." In his other paper, Waksman described the relationship between microorganisms and the carbon/nitrogen ratio of soil.

Although the single, uncategorized program of 1899 was now replaced by eight half days (including simultaneous sessions) on different topics, a strikingly large proportion of papers were concerned with food or water. Under "General Bacteriology," Victoria Carlsson and colleagues warned that "even the cheapest iceboxes should not be sold without thermometers for each shelf . . . to impress upon the user the necessity of low temperatures." Speakers from the National Canners' Association dealt with the heat resistance of spores, while William Albus identified *Clostridium welchii* (now *Clostridium perfringens*) as responsible for the gassy fermentation of niszler cheese.

Under "Agricultural and Industrial Bacteriology," Katherine G. Bitting reported *Bacillus cereus* and other organisms in string beans treated according to published directions for home canning. Other speakers identified capping machines and even pasteurizing equipment as sources of high bacterial counts in milk.

"Comparative Pathology and Immunology" also embraced many food-related topics, including botulism, tuberculosis, and paratyphoid fever. Among the food poisoning outbreaks reported was one in which investigators traced the offending organism to rodent feces on a shelf just above the place where buckets of ice-cream filler were left to cool.

"In all probability these droppings were the source of contamination," the authors concluded. "Probably the cats in the bakery, jumping to this shelf in pursuit of rodents, were the means of brushing some of the droppings into the filler."

The SAB chose the Lord Baltimore and Emerson Hotels in Baltimore, MD, for its 50th (golden jubilee) meeting in 1950. By then, the event had expanded to four full days, with over 250 talks and even more over-lapping sessions (plus four "scientific motion pictures"). Robert Breed and Reuben L. Kahn again spoke, as did a galaxy of other pioneers, including Milislav Demerec, Geoffrey Edsall, Howard Gest, I. C. Gunsalus, Michael Heidelberger, Carl Lindegren, Joseph Melnick, Stuart Mudd, David Pramer, Albert Sabin, Albert Schatz, and Jonas Salk.

Extending the original pool of academics and government scientists, there were speakers from Burroughs Wellcome, Ciba, Pfizer, Sharp & Dohme, Abbott, American Cyanamid, Squibb, and Parke-Davis. Another conspicuous change from 1925 was the number of contributions from army and navy laboratories, including no less than 12 from Fort Detrick, Frederick, MD. There was also a significant growth of non-U.S. partici-pation, with speakers from Havana, Cuba, and from three Canadian cities, plus Aldo Castellani (author of *Microbes, Men and Monarchs*) from Lisbon, Portugal.

One notable link between the 25th and 50th events was the Yale bac-teriologist Charles-Edward Amory Winslow, who spoke on both occa-sions. Delivering the Welch-Novy-Russell Lecture during the Baltimore meeting, he said he believed the honor was due in part to the fact that he was one of only seven surviving charter members. The other reason, he suspected, was that "President Barnett Cohen was the first of my stu-dents at Yale and that he desires to get even with me by calling on me to recite."

Winslow gave a masterly review of historical landmarks, concluding with the work of Martinus Beijerinck and the Delft School and the discoveries of penicillin and streptomycin. He was clearly delighted that his own career had taken him from using the culture tube and microscope slide as the chief tools of research to a point where, following advances in physiology and biochemistry, many papers were now "far beyond my training and experience."

The golden jubilee program confirms the transformation. The subjects of about half of the sessions, ranging from "Sanitation and Soils" to "Dairy and Foods," reflected those addressed 25 and even 50 years earlier, yet the contents of the remainder, on topics such as antibiotics, mutation, electron microscopy, and viruses and rickettsia, would have been well beyond the scope of the previous meetings.

The 1950 Lilly Award winner, Roger Stanier, gave elegant testimony to both the progress and unfinished business of science. He contrasted contemporary knowledge of fermentation, its stepwise metabolism, and the coupling of released energy to anabolic reactions with the understanding of microbial oxidative metabolism. Here, Stanier said, investigators were in roughly the position which Harden and Neuberg occupied 40 years previously as regards glycolysis, with some "interesting and suggestive facts" but no coherent scheme.

"An overall picture of bacterial oxidative metabolism, if one is hardy enough to attempt its construction, strongly resembles a seventeenth century geographer's map of North America. The shape of the continent is roughly outlined, some small local areas already penetrated by the pioneers are correctly shown, but vast regions are either simply left blank or filled in as fancy dictates."

When the (by then) American Society for Microbiology held its diamond jubilee meeting in Chicago, IL, from 12 to 17 May 1974, the program was well into exponential growth. It now encompassed over 1,500 papers, spread over five solid days in the Conrad Hilton Hotel, plus simultaneous sessions in the Blackstone and Pick Congress hotels for much of the time.

Among the speakers were a small number from the 1950 meeting. Especially notable was David Pramer, who in Baltimore had authored a paper on the determination of streptomycin in soil in conjunction with Robert L. Starkey, Waksman's coauthor in 1923. Decades later, in Chicago, Pramer was a coauthor of presentations on the biodegradation of 2,4-dichlorophenoxyacetic acid (2,4-D) and the nuclear polyhedrosis virus of *Porthetria dispar.*

Edwin H. Lennette demonstrated herpesvirus antigen in human brain tissue by solid-phase radioimmunoassay. Joseph Melnick talked on fucosylglycolipids in cells transformed by a temperature-sensitive mutant of murine sarcoma virus, and Eugene A. Delwiche described the effects of polyamines on the growth of *Veillonella alcalescens.*

The diamond jubilee also saw further growth in overseas participation. Noel Carr came from Liverpool University in the United Kingdom to talk about the metabolism of blue-green algae. Henry J. Rogers came from the National Institute for Medical Research in London to discuss bacterial iron metabolism and host resistance. A. C. Matin (later at Stanford University but then at the State University of Groningen in The Netherlands) presented two papers. One was with Hans Veldkamp on freshwater organisms and the other with J. G. (Gijs) Keunen on the active transport of amino acids by membrane vesicles of *Thiobacillus neapolitanus.* Peter Wildy and coworkers from Birmingham Medical School, Birmingham, United Kingdom, authored a paper on herpes simplex viruses.

Among a wealth of other speakers, already famed, of emerging fame, or yet to be widely famed, were Raul Cano, Robert Chanock, Rita Colwell, Roy Curtiss, Stanley Dagley, Arnold Demain, Carleton Gajdusek, Robert Gallo, I. C. Gunsalus, Harlyn Halvorson, Jay Hoofnagle, Edwin Kilbourne, Hilary Koprowski, Ruth Kundsin, Joe Lampen, Thomas Merrigan, Enzo Paoletti, Norman Pace, Stanley Plotkin, Simon Silver, Waclaw Szybalski, Byron Waksman, and David White.

Amidst this panoply of names, and the burgeoning research activities they represented, it seems ironic that the theme for the set piece New Brunswick Lecture, given by Max Tishler on Wednesday of the Chicago meeting, should have been "Is Science Dead?" However, over 30 years later, there is both reassurance and disquietude in the familiarity of three of Tishler's worries: assertions by contemporary critics that science had gotten "too big for its britches," calls for scientists to be "thrown out of the Garden of Eden," and lamentations over "the low esteem of basic research in our national science policy."

While acknowledging that "the exercise of scientific privilege" had brought "staggering benefits," Tishler nevertheless suggested that science may have been "over-selling its beneficence" and "saying too little about the harm it could bring." The way for scientists to retain their privilege, he said, was to "exercise it with responsibility: not responsibility to science alone but primarily to the human condition."

There was a pleasing coherence between Tishler's address and the ASM Lecture, given by René Dubos, with which the diamond jubilee meeting opened. He chose "The Road not Taken," a line from Robert Frost, as the subtitle for his talk on "Pasteur's dilemma." Dubos pointed out Pasteur's awareness, evident from his writings, of the scientific opportunities he had neglected by taking the "pure culture" road in his research. This had been right, though, Dubos argued, because that approach had been an indispensable phase in the emergence of the subject.

"For almost a century," he said, "microbiological sciences have provided models, materials and techniques that have greatly facilitated the understanding of the mechanisms responsible for the operation of isolated biological processes.

"Our science can now . . . provide models, materials and techniques for a new step in the study of life. Just as microbiologists have been pioneers in modern developments of metabolism and genetics, so they can now participate in the effort to understand the integration of individual biological units into more complex social structures—including human societies."

V. Doing Microbiology

56 At the Level of Cowpats

Walk into a physics laboratory these days, and you are unlikely to see anyone working on elementary problems drawn from the everyday world. You will not find experimenters interested in how moving objects impart kinetic energy to stationary ones. There will be nobody investigating the passage of light rays through a convex lens. Studies of this sort have long been supplanted by esoteric scrutiny of subatomic forces and the probing of worlds beyond our own galaxy. Indeed, much of "classical" physics was swept away by the quantum and associated revolutions of the late 19th and early 20th centuries. Modern instrumentation has accelerated the trend.

Other mature sciences show similar pictures. Geologists have long since completed their delineation of the basic structure of the earth's crust. Chemists are no longer adding to the catalogue of elements and their properties and interactions.

Enter a microbiology laboratory, however, and you may well discover white-coated figures poring over problems very similar to those tackled by the pioneers over a century ago. Although many new tools and techniques have appeared in the intervening decades, important questions remain to be solved at the very simplest level. Which organism causes this disease? Which one promotes that natural process? "New" diseases aside, how do some very familiar pathogens get from A to B? The persistence and emergence of unknowns of this sort, despite the transformations spawned by molecular genetics and sophisticated technology, is one of the abiding fascinations of microbiology.

Consider cowpats, for example. I can readily imagine Louis Pasteur peering at these moist brown objects, ruminating over how rainwater carries bacteria away from them, and wondering whether this might pose an infectious hazard. Well, that is exactly what Richard Muirhead of Otago University in Dunedin, New Zealand, has been doing. Working with colleagues at two other New Zealand centers, the Invermay Agricultural College in Mosgiel and the National Institute for Water and Atmospheric Research Ltd. in Hamilton, he has sought to determine in what form and numbers *Escherichia coli* is disseminated from cowpats.

Previous studies demonstrated a clear and predictable correlation between the level of fecal coliforms in cowpats and in runoff water. However, they have left several important questions unanswered, in particular, whether the organisms occur mostly as single cells or are concentrated in flocs or attached to particles of soil. The answers could be highly relevant to the transmission of enteropathogenic strains, such as *E. coli* O157:H7.

Thus, just as Pasteur might have done, members of Muirhead's team visited a farm (every month for just over a year) to collect cowpats. Carefully observing a grazing herd, they leaped into action whenever one of the animals defecated, collecting the material on a shovel and rushing it to the laboratory bench. After taking samples for *E. coli* counts, they then placed each fresh cowpat under an overhead nozzle delivering simulated rainwater; 20 minutes later, they switched off the simulator and collected all the runoff water for analysis.

When they fractionated the water to determine the proportions of bacteria adhering to solid particles, the New Zealanders found that around 92% of the organisms were present as individual cells. Not only were few bacteria attached to particles, but there was also no evidence of flocculation.

"The observation that *E. coli* are eroded and transported from cowpats as individual bacterial cells indicates that they will be highly mobile in overland flow," Muirhead and his colleagues wrote in *Letters in Applied Microbiology* (2). "This apparent lack of a propensity for bacteria to settle or deposit, is also likely to be a cause of the high concentrations of bacteria frequently observed in overland flow from agricultural land, and the sometimes limited attenuation of bacteria through buffer zones."

A few years ago, at a conference in The Netherlands, I read a reassuring poster on this very subject which claimed that the dangers posed by potential pathogens in runoff water on farms had been greatly exaggerated. In reality, the poster said, most organisms would settle with the particulate matter carrying them. Sedimenting into the soil, they would be destroyed within weeks or months by grazing protozoa or some other form of microbial antagonism. The New Zealand work shows that the poster authors, using clever but faulty methodology, were wrong.

Pasteur would have been delighted with this demonstration, not least because an observation of inherent interest concerning microbial behavior could also have far-reaching practical implications. We now know that measures to reduce the level of intestinal bacteria in overland flow on farms need to entrap not only particulate matter, but also single bacterial cells.

A second example of elemental simplicity comes from the field of influenza virus research, which in recent years has focused on the emergence of the H5N1 avian flu and its potential threat to the human population. This example is notable not only for throwing light on these issues, but also for highlighting a deficiency in much conventional epidemiology.

Working at the Karolinska Institute in Stockholm and other Swedish centers, Anna Thorson and her colleagues wondered whether the very high mortality rate reported in Southeast Asia for people contracting the H5N1 virus directly from chickens might be illusory. Was it possible that, in addition to patients sufficiently ill to be admitted to the hospital, there was a wider penumbra of milder cases, many of which were never reported in the formal statistics? If so, the true mortality rate might be much lower than we have imagined.

When the investigators conducted a household survey in a rural area of Vietnam, where some of the first outbreaks of H5N1 flu occurred in chickens and led to fatal human cases, they found evidence consistent with their hypothesis. However, their findings emerged only because they specifically asked people about flu-like illnesses (defined as a cough and fever) that they had experienced around the same time the considerably more serious cases occurred (3).

The implications here were that transmission of the H5N1 virus to people in close contact with chickens may have been more common than hitherto supposed and that the resulting human disease was often less severe than hospital morbidity and mortality data have suggested. In the fullness of time, this verdict, which needs underpinning by experimental work, may prove to be valid or invalid. Its significance, however, is wider. It reminds us that our knowledge of the impact of infectious diseases on human populations comes overwhelmingly from data regarding patients who are sick enough to require hospital treatment. Pasteur would have appreciated the fallacy here.

My last example concerns a process in which humankind has long harnessed microbial skills, accidentally at first but with increasing calculation over the past century: sewage disposal. Bob Seviour, of La Trobe University, Bendigo, Australia, with collaborators at two other Australian centers, has become concerned about our ignorance of what is happening in activated-sludge systems operating throughout the world. For example, though many of these plants include "enhanced phosphorus removal," we have surprisingly sparse knowledge of the organisms responsible.

Information drawn from laboratory or pilot scale systems is not reliable, since the selective pressures in these microbial communities differ from those in full-scale operations. Indeed, Seviour and his coworkers have now demonstrated a clear difference: actinobacteria are probably the predominant phosphorus scavengers in activated-sludge plants, and *Rhodocyclus* is dominant in laboratory scale reactors. Their paper (1) shows how little we still know about the microbial ecology of a process used 24 hours a day to safeguard public health throughout the world.

Cowpats, influenza, and sewage disposal—a visitor from Mars might suppose that earthlings would understand all three as comprehensively as we understand the periodic table or Newton's laws of motion, but microbiology is not like that, is it?

References

1. **Beer, M., H. M. Stratton, P. C. Griffiths, and R. J. Seviour.** 2006. Which are the polyphosphate accumulating organisms in full-scale activated sludge enhanced biological phosphate removal systems in Australia? *J. Appl. Microbiol.* **100:**233–243.

2. **Muirhead, R. W., R. P. Collins, and P. J. Bremer.** 2006. Numbers and transported state of *Escherichia coli* in runoff direct from fresh cowpats under simulated rainfall. *Lett. Appl. Microbiol.* **42:**83–87.

3. **Thorson, A., M. Petzold, T. K. Nguyen, and K. Ekdahl.** 2006. Is exposure to sick or dead poultry associated with flulike illness?: a population-based study from a rural area in Vietnam with outbreaks of highly pathogenic avian influenza. *Arch. Intern. Med.* **166:**119–123.

57 Fishy Business

I (think I) have detected an increase in the frequency of papers appearing in wide-ranging microbiology journals on fish infections and methods of combating them. If reports on piscine pathogens are indeed spilling over from their appropriate specialized publications into the general literature, there is a highly plausible clutch of linked explanations.

First, the traditional toil of hauling fish out of the seas in nets has, like cattle ranching on land, reached its productive limit. Second, this shift has triggered a massive expansion of fish farming, which is now the fastest growing sector of the world's food economy. Having risen by an average of 11% per year over the past decade, aquaculture is expected to overtake beef production as a source of food by 2010. Third, these seismic changes are underlining the pressing need to deal more effectively with the largest causes of financial losses in aquaculture: the molds and bacteria responsible for fish diseases.

Saprolegnia, an oomycete (water mold) related to *Phytophthora infestans,* which precipitated the Irish potato famine at the end of the 19th century, is causing particular concern. As pointed out by Pieter van West of Aberdeen University in Scotland, the dangers posed by this ever-present threat to fish farms have been exacerbated by a decision 6 years ago to prohibit the agent formerly deployed against the organism.

"Up until 2002, *Saprolegnia* infections in aquaculture were kept under control with malachite green, an organic dye that is very efficient in killing the pathogen," van West wrote in *Mycologist* (4). "However, the use of malachite green has been banned worldwide due to its carcinogenic and toxicological effects. . . . This has resulted in a dramatic re-emergence of *Saprolegnia* infections in aquaculture. As a consequence *Saprolegnia parasitica* is now, economically, a very important fish pathogen, especially on catfish, salmon and trout species."

Enthusiasts with fish tanks at home are familiar with *Saprolegnia* as an occasional nuisance in the form of white or grey patches of filamentous mycelium on the body or fins of freshwater species. On a world scale, however, it is a formidable foe. As van West emphasizes, it has not only contributed to crashes in natural populations of salmonids throughout the world, it also causes losses of millions of dollars annually for the aquaculture businesses of the United States, Canada, Scotland, Scandinavia, Chile, and Japan.

In the absence of malachite green, the few alternative chemicals that do staunch the disease in salmonid eggs do not give adequate protection after hatching. Hopes are now pinned on future advances in understanding the basic molecular processes of this ferocious water mold, its interactions with susceptible fish, and the identification of the relevant genes and proteins. It is possible, perhaps likely, that such developments will spawn new strategies to help the ailing aquaculture industry.

Meanwhile, one recent development is already being harnessed in the battle against bacterial fish diseases. This is the use of naturally antagonistic organisms, thought to function either by competitive inhibition, by nonspecifically enhancing immunity, or by modifying the environment. Categorized as probiotics (and associated with a mixture of genuine science and hocus-pocus similar to that which surrounds probiotics in humans), they are already being incorporated into fish feed.

While some of these organisms are lactic acid bacteria, other genera and species are showing a higher degree of genuine efficacy. One recent example is *Bacillus subtilis* AB1, which Brian Austin and coworkers at Heriot-Watt University, Edinburgh, Scotland, and the University of the West Indies in St. Augustine, Trinidad, have developed as a putative agent to combat species of *Aeromonas* that attack rainbow trout.

As described in the *Journal of Applied Microbiology* (3), the organism was one of several obtained from the digestive tracts of euthanized rainbow trout and selected for its inhibitory activity against a pathogenic aeromonad recovered from diseased tilapia. When healthy rainbow trout received *B. subtilis* AB1 in their feed for 14 days, they survived challenge with the pathogen. In immunological assays, the bacillus also specifically stimulated respiratory-burst, serum, and gut peroxidase activities; phagocytic killing; total and alpha-1 antiprotease activities; and lymphocyte populations.

"The fact that *B. subtilis* AB1 was isolated from the gut of apparently healthy rainbow trout confirms the potential role of gut microorganisms in exerting an important role in the wellbeing of the host fish," Austin and his group concluded. The organism "stimulated both cellular and humoral immune responses, which may have provided the rainbow trout with adequate protection to survive the challenge by the highly virulent *Aeromonas* sp." Studies conducted elsewhere have demonstrated that a closely related *B. subtilis* strain shows antibiosis against pathogenic vibrios. It has also been used to improve pond water quality, leading to increased survival of black tiger prawns.

An alternative approach to the control of fish pathogens is, of course, immunization. But the difficulties inherent in its development can be gauged from recent, painstaking progress in studying a genus, *Moritella*, that is taxonomically remote from *Aeromonas* and other bacterial fish pathogens. The psychrophile *Moritella viscosa* is a particular target as the cause of winter ulcers in fish raised in sea cages below 10°C. Endemic in farmed salmonids in North Atlantic countries, the disease causes significant mortality and thus financial losses. Cod is vulnerable, as is turbot, though halibut is relatively resistant.

A polyvalent vaccine has been on the market for several years and has reduced the frequency of the disease. However, there is still considerable room for improvement, and yet, the route toward a more potent and specific vaccine has been bedeviled by an elementary difficulty, the sensitivity of the bacterium to lysis, that reminds us how far fish microbiology has lagged behind other sectors of the subject until comparatively recently. Whereas the fragility of *Neisseria gonorrhoeae* and *Streptococcus pneumoniae* has been recognized for many years, and their autolytic enzymes have therefore been intensively investigated, far less is known concerning those enzymes in marine bacteria.

Two researchers who are now tackling this issue are Eva Benediktsdóttir and Karen Heidarsdóttir at the University of Iceland in Reykjavik. According to a report in *Letters in Applied Microbiology* (1), their work so far should not only facilitate the manufacture of an effective vaccine, but also throw light on the pathogenicity of *M. viscosa*. Their meticulous studies have confirmed the environmental factors responsible for the premature lysis of the bacterial cells and defined the composition of the medium and the temperature that minimize lysis and guarantee successful culture.

Probably the most important single advance in fish pathology over the past year has been the delineation of the complete genome sequence of *Flavobacterium psychrophilum* by Eric Duchaud and colleagues in several different laboratories in France. Causing extensive necrosis known as "cold-water disease" in mature fish and a lethal hemorrhagic septicemia in young fish, *F. psychrophilum* is responsible for considerable economic losses in all major areas of salmonid aquaculture. There is no specific vaccine, and antibiotics do not offer a satisfactory method of control. However, as described in *Nature Biotechnology* (2), the French research is expected to lead to major advances in understanding the molecular pathogenesis of the infection and thus the development of efficient disease control strategies.

Though the tiny crustaceans called sea lice have been in the news recently for their attacks on wild and farmed salmon in British Columbia, it is bacteria, closely followed by oomycetes, such as *Saprolegnia*, that inflict the greatest economic harm on aquaculture worldwide. Notwithstanding progress in biological control and other techniques, genome sequencing seems likely to provide the wherewithal for substantial advances in the future.

References

1. **Benediktsdóttir, E., and K. J. Heidarsdóttir.** 2007. Growth and lysis of the fish pathogen *Moritella viscosa*. *Lett. Appl. Microbiol.* **45:**115–120.

2. **Duchaud, E., M. Boussaha, V. Loux, J. F. Bernardet, C. Michel, B. Kerouault, S. Mondot, P. Nicolas, R. Bossy, C. Caron, P. Bessières, J. F. Gibrat, S. Claverol, F.**

Dumetz, M. Le Hénaff, and A. Benmansour. 2007. Complete genome sequence of the fish pathogen *Flavobacterium psychrophilum*. *Nat. Biotechnol.* **25:**763–769.

3. Newaj-Fyzul, A., A. A. Adesiyun, A. Mutani, A. Ramsubhag, J. Brunt, and B. Austin. 2007. *Bacillus subtilis* AB1 controls *Aeromonas* infection in rainbow trout (*Oncorhynchus mykiss*, Walbaum). *J. Appl. Microbiol.* **103:**1699–1706.

4. van West, P. 2006. *Saprolegnia parasitica*, an oomycete pathogen with a fishy appetite: new challenges for an old problem. *Mycologist* **20:**99–104.

58 Science à la Mode?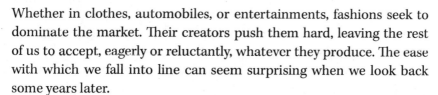

Whether in clothes, automobiles, or entertainments, fashions seek to dominate the market. Their creators push them hard, leaving the rest of us to accept, eagerly or reluctantly, whatever they produce. The ease with which we fall into line can seem surprising when we look back some years later.

None of this has anything to do with science. Modish fancies play no part in an activity that is driven instead by the formulation of evidence-based hypotheses, the design of precision experiments, and the rigorous analysis of results. True, plausible new concepts spread like memes through the international research community, just as ingenious innovations in methodology are taken up promptly by laboratories worldwide, but fashion per se is almost the antithesis of science.

Or is it? If we look carefully at microbiology, or any other scientific discipline, do we not see clear evidence of fashion at work alongside strictly objective modes of thinking? Consider, for example, the idea of quorum sensing. It originated with verifiable facts concerning populations of certain bacteria in which intercellular signals reflect cell density. Over the past decade, however, a more powerful trio of ideas has taken root: that quorum sensing is commonplace, that it has evolved for selective benefit, and that plays a major ecological role in the environment. In reality, however, much of this remains unproven.

The fashionable belief in the widespread significance of quorum sensing is reflected in both textbooks and popular science books and articles. The picture of bacteria as not merely individual, insensitive cells but members of communities, where they sense the presence of their peers and respond in a variety of ways conducive to the common good, is beguiling.

There are, indeed, colorful examples to support this picture. *Photobacterium fischeri*, for example, does not emit blue-green light, or at best produces a dim glow, when the population of cells is relatively scanty. A dense population, on the other hand, generates quite a dazzle. As a sparse community of cells grows, the amount of light does not increase at the same rate. Instead, there is a population size, a quorum, above which all of the bacteria begin to emit much more light than they did before.

Whether in a scanty or dense community, *P. fischeri* produces an autoinducer that can switch on genes in the bacterium that make it luminesce. However, the autoinducer has to reach a particular level before it works. When that concentration is attained, by the right quorum of cells, then the lights go on throughout the population.

Quorum sensing by *P. fischeri* seems to reflect its alternative lifestyles. It can grow freely in the sea, where its population density is very low and where it does not luminesce. However, evolution has also led to an arrangement in which the light organs of certain squids and fishes provide a home for the bacterium. When *P. fischeri* takes up residence in a young fish, it proliferates quickly, producing autoinducer, which soon accumulates to the level at which the entire community begins to generate light.

A few pathogens have also attracted interest through their intercellular signals, not least because of the prospect of thwarting microbial communication as a form of therapy. It might prove possible to combat opportunistic infections caused by *Pseudomonas aeruginosa*, for example, by interfering with its synthesis of, or response to, N-acetyl-L-homoserine lactone (AHL). Two genes whose expression is mediated by this intercellular signaling metabolite are those encoding elastase production in *P. aeruginosa* and cellulase formation in the plant pathogen *Erwinia carotovora*.

However, major issues remain to be resolved. As Mike Manefield and Sarah Turner of CEH Oxford in Oxford, United Kingdom, have pointed out, fundamental questions about the ecology and evolution of gene expression mediated by AHL are just beginning to be asked. "Only one hypothesis (quorum sensing) describing the selective advantage of AHL-mediated gene expression is widely known," they wrote in *Microbiology* (1). Moreover, "the quorum sensing hypothesis has never been tested in the environment. Indeed, the role of the AHL intercellular signaling metabolite in mediating interactions between bacteria and their biotic and abiotic environment globally is poorly understood, if ever considered."

Another lacuna concerns the taxonomic distribution of bacterial signaling. "Statements suggesting that AHL-mediated gene expression is a widespread phenomenon in Gram-negative bacteria often appear in reviews or introductions on the subject (and certainly in grant applications)," wrote Manefield and Turner. "Yet a simple detailed survey of its distribution amongst bacterial taxa has never appeared in the literature."

Data presented by the Oxford duo indicate that known AHL-producing bacteria are actually rather uncommon. "What you don't often hear is that within the alpha *Proteobacteria* only 7 out of 133 (5.3%) genera harbor AHL-producing species," the authors wrote. The corresponding figures for beta- and gammaproteobacteria are 4 out of 56 (7.1%) and 10 out of 180 (5.5%), respectively. These data "reflect the fact that AHL production is restricted to a limited number of *Proteobacteria*, let alone Gram-negative bacteria, and should therefore not be considered widespread."

Much fuller knowledge of the taxonomic distribution of the phenomenon is, of course, only the first step toward real comprehension of its ecological significance in the biosphere. For the moment, it seems premature to suggest that we really understand either.

We might benefit, too, by looking back to another occasion when a modish idea was grasped a little too avidly. The person who argued thus was the late N. W. "Bill" Pirie of Rothamsted Experimental Station in Harpenden, United Kingdom. A doughty critic of the uncritical acceptance of new notions, Pirie tackled one such example in a provocative yet astute article entitled "Biological replication considered in the general context of scientific illusion" in *New Biology* (2).

"Man is a myth-making animal; scientists belong to the category man; so scientists are myth-making animals," Pirie began. "That syllogism is in accordance with the principles of formal logic and with experience." He then argued that, after decades when proteins were considered to carry hereditary characteristics, scientists had been overhasty in switching their allegiance to nucleic acids.

Pirie gave as one example Alfred Hershey's work on the movement of components of the T2 bacteriophage into *Escherichia coli*. "He finds that only 1% of the protein is transferred. That seems very little to those who do not remember how large T2 phage is compared to many other viruses; the weight of 1% of the particle is 1.2Md, and this is the weight commonly attributed to the infective fragments of TMV."

Pirie reasoned that this quantity of material could not be regarded as negligible. The technique used to produce it was not intended to produce a complete balance sheet but simply to search for particular categories of substance. Indeed, Hershey himself pointed out that, although protein and some other materials were transferred along with DNA, he had chosen to focus on DNA as a working hypothesis.

"This careful separation of observation from assumption has not been noticed by those in the rout or claque that follows so noisily behind," Pirie wrote. "They assert boldly that DNA is the only material transferred to the infected cell and thus turn a hypothesis into a dogma."

It is now 5 years since we celebrated the 50th anniversary of the discovery that the structure of DNA does account for its function as the carrier of genetic information. Nevertheless, we should remember that some of the evidence was less than conclusive when it was first presented. Another example of fashion at work?

References

1. **Manefield, M., and S. L. Turner.** 2002. Quorum sensing in context: out of molecular biology and into microbial ecology. *Microbiology* **148:**3762–3764.

2. **Pirie, N. W.** 1960. Biological replication considered in the general context of scientific illusion. *New Biol.* **31:**117–135.

Animalcules: the Activities, Impacts, and Investigators of Microbes

59 "Wherever They Are Found…"

The declared purpose of the Institute of Microbiology, Rutgers, NJ, at its dedication in 1954 was "the study of the smallest forms of life, the microbes, wherever they are found and no matter what their activities may be." That philosophy was embodied in the life of the institute's inspirational first director, Selman Waksman. Working with microorganisms in health and disease, in the soil and elsewhere, Waksman saw little point in boundaries that divided microbiology into two or more different subdisciplines. He simply ignored them.

Over half a century later, it is time to listen afresh to Waksman's message and to consider the need for a more unified approach to the microbial world. The thought is prompted both by the general advances in molecular genetics and microbial ecology that surround us on all sides these days and by three specific developments.

These are the discovery of high levels of antibiotic resistance in plant-colonizing enterococci; the challenge of dealing with an organism such as *Burkholderia cepacia*, which causes both plant and human infections; and the recognition of protozoa as reservoirs of animal (including human) pathogens. One might also include the United Kingdom outbreak of bovine spongiform encephalopathy in the late 1990s, together with the related human variant, Creutzfeldt-Jakob disease. The handling of this episode was scarcely helped by an unfortunate demarcation between agricultural and medical microbiologists.

Concern about enterococci has, of course, been growing for some time, as they have increasingly been recovered from urinary tract and other opportunistic and nosocomial infections (1). One species, *Enterococcus faecium*, is characterized by particularly high levels of antibiotic resistance, while enterococci in general can usually transfer their resistance genes quickly to closely related bacteria and to disparate genera, as well.

Not surprisingly, these findings have emerged from studies on isolates from humans and other animals, whose intestines they inhabit. There has also been a large amount of work on enterococci in municipal wastewaters (where they are significant indicators of fecal pollution).

In contrast, enterococcal colonization of plants has attracted serious interest only very recently. Vaguely recognized but largely neglected, phylloplane-colonizing members of the genus *Enterococcus* nevertheless appear to threaten human health. Apparently far commoner than was thought hitherto, they are often insensitive to commonly used antibiotics. In 2001, at least one new species was reported, and there may well be more to come.

One group pioneering the investigation of plant-associated enterococci is based at the University of Rostock and other centers in Germany. The team set out to determine their prevalence on samples of grass taken from the pristine environment of meadows in northeast Germany that were surrounded by grassland and where no manure had been used as fertilizer.

The results (4) were rather worrying. First, they showed that enterococci were extremely widespread. Secondly, while strains of *E. faecium*, *Enterococcus faecalis*, and other species were recovered, most isolates belonged to a new genotype (as characterized by techniques including restriction analyses of PCR-amplified 16S rRNA genes). This probably represents a formerly unknown species that has evolved in the harsh environment of leaf surfaces.

Antibiotic sensitivity tests on 204 isolates showed that virtually all (94% and 98%, respectively) were insensitive to gentamicin and streptomycin; 39% were unaffected by rifampin, 12% by penicillin G, and 7%

by erythromycin. Although there was no evidence of vancomycin resistance, these levels of potentially transferable resistance are disquieting.

From a broad microbiological perspective, one thing is especially odd about these discoveries. It is now 23 years since enterococci were given a new systematic position, when the former *Streptococcus faecium* and *Streptococcus faecalis* were transferred to the genus *Enterococcus*. Until recently, however, there has been no dedicated study of plant-associated enterococci. While the literature contains sporadic reports of unidentified streptococci/enterococci in plants before and after 1984, their possible relevance for human health has received virtually no attention whatever. Could the dissociation between medical, plant, and environmental microbiology be responsible?

Much the same question might be asked concerning *Burkholderia cepacia*. Agricultural microbiologists, on the one hand, have studied this organism as the cause of diseases such as bacterial rot of onions and, more recently, as a potentially highly effective agent to combat other plant infections. It has also turned up as a contaminant of water supplies on the space shuttle and has become notorious for its metabolic versatility, which includes the capacity to exist on penicillin G as a sole source of carbon.

Medical microbiologists, on the other hand, know *B. cepacia* as an opportunistic pathogen capable of chronically colonizing the lungs of cystic fibrosis sufferers. It can also cause serious problems for patients with chronic granulomatous disease and in those with life-threatening lung infections who require artificial ventilation.

These two facets of *B. cepacia* clearly come into some degree of conflict as regards recent proposals to use the organism as a biological control agent. However, as John Govan and Peter Vandamme pointed out some years ago (2), these aspects have been investigated without the degree of collaboration that might well have been mutually beneficial, not least as regards risk assessment.

"What better microbial challenge to unite agricultural and medical microbiologists," they wrote, "than an organism that reduces an onion to a macerated pulp, protects other crops from bacterial and fungal diseases, devastates the health and social life of CF patients, and not only is resistant to the most famous of antibiotics, penicillin, but can use it as a nutrient?" Yet a perusal of the literature over the years since those words were written indicates that much work on different aspects of this organism is proceeding in relative isolation.

My third example concerns the significance of protozoa as reservoirs for pathogens, not only the comparatively well-understood case of *Legionella pneumophila* (3), but other bacteria too (see chapter 27). It seems that residence inside protozoa can both protect organisms from adverse environmental conditions and enhance their infectivity. One analogy is that they use their hosts as "biological gymnasia" where they can "train" for encounters with more evolved mammalian cells. Protozoa, in other words, may be sources of emerging pathogens, helping bacteria to make the transition from the environment to human and other animal hosts.

As evidence grows to support the practical importance of this scenario, the need for a less fragmented approach to microbiology becomes correspondingly more apparent. So, too, with my other examples. This means not only greater collaboration between different species of microbiologist, but also a more unified approach in teaching the subject.

Waksman would have been delighted, too, over the way in which molecular genetics has transformed and deepened our comprehension of microbial activities in recent decades, reinforcing the need for collaboration between the various sectors of the subject he did so much to establish. Tim Stinear of Monash University in Clayton and Paul Johnson of Austin Hospital, Heidelberg, both in Victoria, Australia, gave a stunning example of this dimension when they described (5) how lateral gene transfers and genome reduction had apparently remodeled a fish pathogen into the agent that causes Buruli ulcer in humans.

Whether considering animalcules genotypically or phenotypically, Waksman was right.

References

1. **Edwards, D. D.** 2000. Enterococci attract attention of concerned microbiologists. *ASM News* **66:**540–545.

2. **Govan, J. R., and P. Vandamme.** 1998. Agricultural and medical microbiology: a time for bridging gaps. *Microbiology* **144:** 2373–2375.

3. **Harb, O. S., and Y. Abu Kwaik.** 2000. Interaction of *Legionella pneumophila* with protozoa provides lessons. *ASM News* **66:**609–616.

4. **Müller, T., A. Ulrich, E. M. Ott, and M. Müller.** 2001. Identification of plant-associated enterococci. *J. Appl. Microbiol.* **91:**268–278.

5. **Stinear, T., and P. D. R. Johnson.** 2007. From marinum to ulcerans: a mycobacterial human pathogen emerges. *Microbe* **2:**187–194.

60 There's More To Do

Microbiologists may entertain a degree of skepticism toward books that have portrayed human society as virtually helpless in the face of viral evolution and opportunism. The 2002 International Union of Microbiological Societies (IUMS) World Congress provided a vigorous corrective to such skepticism. Held in the Palais des Congrès in Paris, the event was remarkable in providing a plethora of evidence about hitherto-unrecognized viruses in humans and other animals and plants and about new behaviors and geographical incursions of familiar pathogens.

Particularly disquieting was news of the emergence of an African mosquito-borne flavivirus, a member of the Japanese encephalitis group, in central Europe. As described by Herbert Weissenbock and colleagues at the Institute of Pathology and Forensic Veterinary Medicine in Vienna, Austria, the story began in 2001 when observers in that country reported a series of deaths in different species of birds, recalling the commencement of the West Nile virus epidemic in New York in 1999.

In the later case, sequencing and phylogenetic analysis of viral isolates showed them to be 97% identical to Usutu virus (USUV). A relatively unknown member of the genus *Flavivirus*, USUV is closely related, not only to West Nile virus, but also to other important human pathogens, such as those causing Japanese encephalitis, Murray Valley encephalitis, and yellow fever. It had never previously been described outside Africa.

Weissenbock and his coworkers reported that the organism was highly pathogenic for several different species of birds, though they were unsure whether it had the capacity to cause severe human disease. In a paper in *Emerging Infectious Diseases* (1), they called for the establishment in Europe of surveillance programs for mosquito-borne flaviviruses (based on virus detection in mosquitoes and dead birds, as well as epidemiological studies) like those that were initiated in the United States after the first detection of West Nile virus.

Reports from Novosibirsk, Russia, and Eilat, Israel, provided evidence about two related threats. The Russian investigators described fatal cases of tick-borne encephalitis in their region attributed to a new variant, which also caused a hemorrhagic syndrome, until then not known to be associated with the disease. Also, the trapping and sampling of palearctic birds migrating through Israel during the spring and fall in recent years has shown higher levels of West Nile virus infection than were indicated by previous studies. This finding may help to account for the recent reemergence of epizootics of this disease in humans, horses, and domestic geese in Israel and several European countries.

IUMS congress delegates also heard of the first success of a Europe-wide network established to facilitate the investigation of the sources and spread of gastroenteritis caused by Norwalk-like caliciviruses. An outbreak in France was traced to drinking water but then spread to three other countries through imported French oysters. The agent proved to be a novel variant, designated 11b, which was eventually responsible for 109 outbreaks in eight different countries.

An entirely new virus, reported in Paris by Ian Mackay from the University of Queensland in Australia was a member of the family *Paramyxoviridae*. Termed human metapneumovirus, it appeared to be responsible for some of the third or so of respiratory tract infections that otherwise yield no causative agent. Mackay and his collaborators recovered human metapneumovirus from children with mild-to-severe respiratory tract disease in several Australian hospitals. They reported that it showed a high degree of homology with isolates reported around the same time in The Netherlands.

Two reports highlighted possible underestimates of the capacity of pathogens to thwart host defenses through antigenic changes. Contrary to the belief that recombination in hepatitis C virus was rare in vivo and that recombinants were usually not viable anyway, research in Sweden and Russia had demonstrated that natural recombinants do exist and may play a role in creating genetic diversity in the organism. Other contributors to the congress warned that strain diversity among rotaviruses is much greater than was previously believed, while serotypes formerly considered to be very uncommon are important causes of diarrhea.

From the University of São Paulo came evidence that a new P genotype rotavirus was circulating in the pig farms of southern Brazil. Other animals reported to be harboring previously unknown organisms included walruses and mink. A paper from the Centers for Disease Control and Prevention in Atlanta, GA, described the isolation and characterization of the new walrus calicivirus, while investigations on Swedish and Danish mink farms had identified a novel astrovirus that might be responsible for much of the preweaning diarrhea that causes considerable economic losses in Europe.

Announcements of hitherto-unrecognized plant viruses were, if anything, more numerous in the Palais des Congrès than those of human and other animal pathogens. From California and Arizona came news of a new soilborne tombusvirus causing necrosis and dieback in lettuce. Italian researchers described an entire new genus of viruses affecting the grapevine, and there were fears regarding the emergence and recombination of single-stranded DNA geminiviruses, which are slashing yields of crops, such as tomato, cotton, and cassava, in Africa, Asia, and the Mediterranean region.

The IUMS is composed not only of a division of virology, but also of divisions devoted to other categories of microbial life. It was pleasing, therefore, to find that some of the most arresting topics considered in Paris were at the interfaces between different disciplines.

One instance was the work of Jean Patterson and her coworkers at the Southwest Foundation for Biomedical Research in San Antonio, TX. They had been studying the 5.3-kb double-stranded RNA virus that occurs in several strains of the parasitic protozoon *Leishmania* and had developed an assay for the virus based on a highly conserved untranslated region at the 5' end of its genome. When they used the assay on samples from patients with leishmaniasis, they recorded the most positive results in individuals who had cutaneous lesions and scars. This finding provided the first clue to the role that the virus plays in the pathogenesis of the disease.

An analogous question lurked behind the presentation of Anne Lanois and Alain Givaudan of the University of Montpellier in France. Their work centered on *Steinernematidae* nematodes, which are pathogenic in various insects. Inside their intestinal vesicles, the nematodes harbor symbiotic bacteria, principally the gram-negative *Xenorhabdus nematophila*. What is its role when carried by the worm into its insect prey? Part of the answer came from the French group's demonstration that genetic modification to deprive *Xenorhabdus* of its flagellae also abolished the secretion of its extracellular hemolysin. This, in turn, attenuated the nematode's virulence in insects.

There were, of course, many items of good news, as well as bad, in the Palais des Congrès. These included announcements about novel organisms showing promise as biopesticides. One of these was a densovirus infecting green peach aphids, which was described by Plant Research International at Wageningen in The Netherlands. There were pointers, from the Max von Pettenkofer Institute in Munich, Germany, toward immunoprophylaxis for prion diseases, and there was news from USAMRIID, Fort Detrick, MD, of a potent vaccine candidate that protects mice against highly virulent Ebola virus.

Still, the strongest impression was one of continual challenge. That challenge stems from the unique capacity of microorganisms to reshape their genotypes, their phenotypes, and their environments, which include us and all of the other elements of the biosphere upon which our lives depend. For microbiologists, there will always be work to do.

Reference

1. **Weissenböck, H., J. Kolodziejek, A. Url, H. Lussy, B. Rebel-Bauder, and N. Nowotny.** 2002. Emergence of Usutu virus, an African mosquito-borne flavivirus of the Japanese encephalitis virus group, in central Europe. *Emerg. Infect. Dis.* **8:**652–656.

61 Self-Frustration

During World War II, the personal scientific adviser to Winston Churchill, the British prime minister, was Lord Cherwell. Appropriately to his high status, Cherwell carried a special security pass, intended to give him immediate access to any government establishment around the country. Unlike the passes issued to more numerous officials of lower rank, his type was limited to a tiny handful of exceedingly important people. Moreover, its appearance was kept confidential so that it could not be forged. In consequence, Lord Cherwell had frequent arguments with guards and doormen who, never having seen such a pass before, refused to let its angry holder enter their premises.

Another government scientist, R. V. Jones, recounted this example of what he described as self-frustration in *Nature* (4), one of several wartime tactics that had the very opposite effect of that which was intended. The story came to my mind in an entirely different context, when I read some comments by Cecil H. Fox of Molecular Histology Inc. in Montgomery Village, MD. Writing in *Science* (2), Fox argued that high-containment P4 facilities, while playing an important role in our defense against communicable disease, also bring certain dangers.

In addition to the well-known P4 laboratories at the Centers for Disease Control in Atlanta, GA, Porton Down in the United Kingdom, and Novosibirsk in Russia, Fox used the World Wide Web to discover

Animalcules: the Activities, Impacts, and Investigators of Microbes

a total of 24 or so throughout the world, some of them then still under construction. He suggested that such facilities should not be seen as objects of prestige, like the national airlines of developing countries.

"If the public believes that P4 facilities are their first line of defense for new or emerging illness, then the public must also be assured that their safety is at least as rigidly controlled as is that of the nuclear reactor industry," Fox argued. "A spill in a poorly managed P4 facility could be far more devastating than Chernobyl or Three Mile Island."

P4 laboratories have attracted increasing interest because of the perceived threat of bioterrorism. They could, of course, do little or nothing to prevent such attacks from taking place. However, Fox went further than this, asserting that laboratories of this sort would be very little help even afterwards. In addition, "P4 facilities provide a wonderful setting for producing potential weapons and should be considered with the same regard as nuclear weapon storage facilities."

Fox's view that P4 facilities would be of little value in responding to a bioterrorist incident is questionable. They could be at the very forefront of investigation and countermeasures. One may be skeptical, too, about the practicality of Fox's further suggestion that equipment for furnishing such facilities should be regulated as stringently as the equipment for plutonium technology. Nevertheless, there was more than a grain of truth in his analysis.

Indeed, the evolution of microbiology over the last century has been accompanied by many more examples of what R. V. Jones calls self-frustration. Consider the strategies we have deployed to vanquish pathogenic bacteria. Antimicrobial drugs have saved countless lives but have also created the ominous specter of multiply resistant superbugs.

Similarly, hospitals have helped to win many struggles against communicable disease but have also become prime locations for patients to acquire virtually untreatable infections. Replacing small local hospitals with larger centralized ones, in order to focus high-tech facilities and provide better care, has extended further the opportunities for "hospital strains" to proliferate and cause more opportunistic infections.

Antibiotics, killing nonpathogens as well as pathogens, have made candidiasis a more serious problem than it was before. The eradication of smallpox has turned the remaining stocks of variola virus into one of the most dangerous materials in the world, and the increased use of refrigeration and chill compartments in supermarkets, positive developments in the drive to minimize food-borne infection, have turned *Listeria*, which thrives at those temperatures, into a more insidious and formidable foe than it was hitherto.

One of the oddest examples of self-frustration concerns the viruses, bacteria, fungal spores, and protozoan trophozoites and cysts that can stick avidly to clothing. Decades ago, the piping hot temperatures used for the weekly wash at home, and in commercial laundries, were sufficient to sterilize garments of all but the most resistant spores. Then came two entirely laudable ecological concerns.

First, growing anxiety about energy saving triggered a shift away from washing with scalded fingers toward much lower temperatures, which are tolerated by most microorganisms. Second, environmental researchers traced the eutrophication of rivers and lakes to salts in the water discharged into these natural water systems. A key culprit proved to be sodium tripolyphosphate, which in consequence was removed from commercial laundry detergents, or at least drastically reduced in concentration.

However, the main role of sodium tripolyphosphate had been to prevent organic and inorganic substances from accumulating on fabrics. Washed repeatedly without it, garments accumulate nasty, tenacious, insoluble encrustations. Have we exacerbated one problem through well-intentioned measures to deal with another?

When Carlotta Granucci and colleagues at the University of Milan, Milan, Italy, studied the microbiological dimension of the encrustation, their findings (3) were not encouraging. They laundered pieces of cotton, terrycloth tea towels, and a cotton-and-polyester fiber 25 times in a household washing machine with various brands of detergent. Afterward, they measured the amounts of organic and inorganic encrustation and agitated strips of material in nutrient medium to see what proportion of bacteria were released.

The results showed that, while small percentages were eluted, the majority of organisms remained firmly stuck. The degree varied according to the fabric, but encrustation clearly enhanced the adhesion of bacteria, the two phenomena being directly proportional to each other.

Self-frustration has characterized the development of microbiology as a practical science, too. One person who pointed this out back in 1950, and was regarded as a dangerous iconoclast for doing so, was Kenneth Bisset of the University of Birmingham, Birmingham, United Kingdom. However, it is now clear that his analysis was largely correct.

Writing in *The Cytology and Life-History of Bacteria* (1), Bisset pointed out that many of the techniques then used to study bacteria had "tended to obscure rather than clarify underlying truths. Especially is this true of the staining techniques employed for routine examination of bacteriological material and cultures. The distorted vestiges of bacteria which survived the technique of drying and heat-fixation were accepted as truly indicative of the morphology of the living organisms." In fact, these methods had positively impeded the discovery of intracellular structures, which after more than half a century of bacteriology were only then beginning to be revealed.

Much the same could be said of efforts to recover and characterize specific organisms from infections and locations such as the soil. Plating and selective culture techniques were undoubtedly cornerstones for the development of both medical and non-medical microbiology, yet they also encouraged styles of thinking, reflecting growing understanding of how microorganisms behave in pure culture, that were founded upon artificiality. Only in much more recent years have we been garnering substantial knowledge about the real microbial world of heterogeneous communities, consortia, and biofilms (see chapters 15 and 16).

It seems implausible that all of the self-frustrations of microbiology lie in the past and inherently plausible that further examples are embedded in the science today. I wonder what they are?

References

1. **Bisset, K.** 1950. *The Cytology and Life-History of Bacteria.* E&S Livingstone, Edinburgh, United Kingdom.

2. **Fox, C. H.** 1999. Hot zones. *Science* **284:**261.

3. **Ghione, M., D. Parrello, and C. Granucci.** 1989. Adherence of bacterial spores to encrusted fibres. *J. Appl. Bacteriol.* **67:**371–376.

4. **Jones, R. V.** 1965. Impotence and achievement in physics and technology. *Nature* **207:**120–127.

62 Genomics and Innovation in Antibiotics

Newspapers are more frequently and plausibly blamed for sensation-alism than scientific journals, yet it is from journals that I have, over the past decade, collected many articles with titles such as "Mining the Genome for Drugs" and "The New World of Designer Drugs." These two (from *Science* and the *British Medical Journal*, respectively) summarize a compelling syllogism. It goes as follows.

Every newly sequenced genome reveals hitherto-unknown proteins whose roles can then be determined. Secondly, chemotherapeutic agents can be developed to interfere with or enhance those functions. Therefore, medicine is now entering a new golden age in which phar-macology, formerly based on empiricism and serendipity, will yield rich hauls of specific, potent, rationally designed treatments for disease.

The argument is applied with particular enthusiasm to interven-tions directed toward single-nucleotide polymorphisms in the human genome, but it is also deployed in the case of new targets, potentially vulnerable to antibiotic attack, which are embedded in microbial gene sequences.

There seems little doubt that important advances in both sectors will emerge from this modish strategy. At the same time, it may be pertinent to ask whether such rewards will match the hype as quickly as some have supposed, if, indeed, ever.

A closely coupled concern is whether the headlong rush into genomics has unreasonably marginalized existing approaches to pharmaceutical innovation. This could be true in the antimicrobial arena, where bacterial sequencing appears to have arrived just on cue to help us cope with the twin challenges of diminishing returns from traditional strategies and the specter of untreatable infections caused by multiply resistant pathogens.

Has the genomics revolution persuaded some researchers that earlier techniques can now be consigned to history? Has the search for novel bioactive compounds in nature been reduced prematurely? Are we overlooking the mass of knowledge we already possess on organisms known to produce antibiotics, previously found to be unsatisfactory for one reason or another but which might now be reexamined in light of modern knowledge (including genomics)?

One person who has donned the garb of Cassandra to examine the impact of gene sequencing on therapeutic innovation in general is David Horrobin, who was chief executive of Laxdale Research in Stirling, Scotland. Writing in *Nature Biotechnology* (1), he observed that, "with rare exceptions, most of the top 20 multinational pharmaceutical companies are not generating in-house the new products needed to sustain the rates of growth they have enjoyed in the past."

The promise is that genomics, together with combinatorial chemistry and high-throughput screening, will bring relief from that grim scenario. Will it happen? "Yes, certainly," was Horrobin's answer—with a crucial proviso. "It may not occur within the commercial lifetime of most existing companies. Within 20 years, almost none of today's big pharmaceutical or biotechnology companies will exist in the form in which they exist today. Few of the investors and shareholders who have poured so much money into the new technologies will see any substantial reward."

Horrobin was talking primarily of possible therapies for human gene-related diseases, where there are certainly grounds to support his caution. For example, many such conditions involve two or three different genes. The need for a corresponding number of different drugs, each with its many hurdles of toxicology testing and clinical trials, indicates that reality is far away from the "quick fix" suggested by certain protagonists.

As regards the hunt for novel antibacterial agents, we have certainly come a long way over the past decade in compiling blueprints for action. Completed genome sequences, from *Haemophilus influenzae* in 1995 to *Sorangium cellulosum* in 2008, not only reveal many hitherto-unrecognized targets for attack, but also emphasize how limited are the numbers of bacterial functions that we thwart with conventional antibiotics.

Data of this sort will clearly help us, and quickly, to characterize many targets. Comparative genomics will assist further in eliminating, as possible candidates, genes which are functional in other organisms but whose activities are no longer required by the pathogen in question.

Even when we are armed with information of this sort, however, the long journey of drug development remains as before. There is still the challenge of finding (small-molecule) agents to inhibit these targets or their functions. Suitable natural or synthetic compounds remain to be discovered, just as chemical manipulations will be required in most cases to turn them into actual drugs.

Meanwhile, much surely remains to be gained from the corpus of knowledge (and of bacterial strains) accumulated during decades of conventional research on antibiotics. Several potential avenues were highlighted by Hans Zaehner and Hans-Peter Fiedler of the University of Tübingen, Tübingen, Germany, in *Fifty Years of Antimicrobials* (4).

One idea is the reinvestigation of antimicrobials, which have already been introduced into niche markets but which, in part because of the dominance of beta-lactams, have not been optimized to anything like the same degree. Examples are cycloserine, viomycin, fumagillin, albomycin, and coumermycin.

Secondly, it seems highly likely that among the 99% of over 4,000 known antibiotics rejected years ago for human treatment there are many candidates that could now be optimized for medical use. Among such agents (also discarded because of the appeal of beta-lactams) are orthosomycins, everninomycins, avilamycins, euramycins, curamycins, and lankacidins.

Zaehner and Fiedler made many other proposals, including searches for new antibiotics using novel test methods, examination of previously unculturable bacteria, innovative culture conditions, and means of overcoming penetration barriers to render drugs effective against

Animalcules: the Activities, Impacts, and Investigators of Microbes

organisms they do not at present touch. None of this advice was particularly radical, yet an examination of the literature does not suggest that these avenues are being explored with due vigor.

Given the great rewards of soil screening in the past, we should also remember that remarkable organisms continue to be discovered "out there," though usually by ecologists rather than by biotechnologists interested in antibiotic production. Since well under 1/10 of the world's 1.5 million fungi have yet been characterized, there is much more to come.

Some years ago, Alan Paton and his colleagues in Aberdeen, Scotland, initiated some research on associations between plant cells and bacterial L-forms. One of the early papers from his group (3) showed that L-forms living in close association with Chinese cabbage cells could make the plants resistant to bacterial infection. Aside from their inherent fascination, the consortia appeared to have practical potential in plant protection and indeed showed some promise in that direction (2).

How, I asked Paton, did he first discover this phenomenon? "Simply by spending hours and hours doing what microbe hunters don't do any more—looking down a microscope," he replied. Paton regretted that the advent of gene cloning and sophisticated molecular genetics had led many bacteriologists to abandon some of their traditional techniques entirely.

There is a lesson here for other sectors of microbiology.

References

1. **Horrobin, D. F.** 2001. Realism in drug discovery—could Cassandra be right? *Nat. Biotechnol.* **19:**1099–1100.

2. **Innes, C. M. J., and E. J. Allan.** 2001. Induction, growth and antibiotic production of *Streptomyces viridifaciens* L-form bacteria. *J. Appl. Microbiol.* **90:**301–308.

3. **Waterhouse, R. N., H. Buhariwalla, D. Bourn, E. J. Rattray, and L. A. Glover.** 1996. CCD detection of lux-marked *Pseudomonas syringae* pv. *phaseolicola* L-forms associated with Chinese cabbage and the resulting disease protection against *Xanthomonas campestris*. *Lett. Appl. Microbiol.* **22:**262–266.

4. **Zaehner, H., and H.-P. Fiedler.** 1995. The need for new antibiotics: possible ways forward, p. 67–84. *In* P. A. Hunter, G. K. Darby, and N. J. Russell (ed.), *Fifty Years of Antimicrobials*. Cambridge University Press, Cambridge, United Kingdom.

63 The Relevance
of Taxonomy

Of all the branches of microbiology I first encountered as a student some 50 years ago, taxonomy was the one that seemed seriously boring. Virology was compelling because it dealt with the great plagues of world history and the new threats of today, while protozoology concerned the scourges of the tropics and their fascinating life cycles. Mycology dealt with the antibiotic-producing superbugs and soil bacteriology with the very maintenance of life on Earth. By comparison, taxonomy was a tedious art, practiced by the equally tedious. It was an abstract art with no apparent relevance to the worlds of applied microbiology.

I hope today's students take a more enlightened view. There are several reasons why they should do so: the need for unambiguous definition and discrimination of microbial strains for patent purposes, for example. Another aspect of the practical importance of taxonomy was sharply highlighted by a paper in *Microbiology* (2). It contained a chemotaxonomic and molecular biological comparison of various isolates of the extremely familiar *Pseudomonas aeruginosa*. At first sight, this was not a topic to stir the imagination or stimulate the intellect. Nevertheless, it was an absorbing and to some degree worrying paper, which illustrated the crucial importance of taxonomy in assessing the safety of various modern and future applications of microbiology.

The report came from an avenue of research initiated back in the 1970s by Donald Westlake and colleagues at the University of Alberta in Edmonton, Canada. This latest chapter was written in association with Harry Ridgway, who works for the Orange County Water District in Fountain Valley, CA, and Julia Foght at the Laboratory Centre for Disease Control, Ottawa, Canada. Their principal finding, that strains of *P. aeruginosa* obtained from clinical infections were indistinguishable from those that are prime candidates for the bioremediation of oil-contaminated soil and water, was not only inherently interesting, it was also significantly disturbing.

Westlake first became interested in pseudomonads as hydrocarbon degraders when he recovered a number of different species with this ability from soil that for several years had been soaked with diesel fuel. This led him to investigate the effect of factors such as the incubation temperature and the composition of crude oils on their biodegradation. Subsequent studies of water and sediments from the Straits of Juan de Fuca in Washington State also showed that even sites where oil had not been spilled usually contained oil-degrading populations in which *Pseudomonas* species were prominent.

In 1990, Harry Ridgway complemented Westlake's work when he published an account of bacteria isolated from a shallow coastal aquifer contaminated with unleaded gasoline. Analysis of protein banding patterns, together with conventional phenotypic tests, established that 87% of 244 organisms recovered from the site belonged to the genus *Pseudomonas*. Of these, 89 isolates proved to be strains of *P. aeruginosa*. All degraded naphthalene, *p*-xylene, ethyl benzene, and other low-molecular-weight aromatics. They fell into two distinct taxonomic clusters, and members of one group from each also grew on alkanes, such as *n*-hexane and methyl-cyclopentane.

More discriminating tests conducted before and after bioremediation of the aquifer threw light on another species, too: *Pseudomonas mendocina*. Analyzing genomic DNA by means of restriction fragment length polymorphisms, in addition to protein banding, Ridgway and collaborators discovered that prior to the commencement of bioremediation, most of the hydrocarbon-degrading capacity of the bacterial flora occurred in strains of *P. aeruginosa*. However, once the cleansing process was under way, *P. mendocina* played the dominant role in attacking the pollutants.

However, what of the clinical significance of these organisms? *P. aeruginosa* is a well-known opportunistic pathogen, particularly in nosocomial and pulmonary infections in patients with cystic fibrosis (CF). Several discoveries have provided insight into these dangers. One group (1) recovered *P. mendocina* from a patient with infective endocarditis. Another (3) reported that at least some strains of *P. aeruginosa* have pathogenic potential in both animals and plants. A third team (4) discovered that the most frequently identified *P. aeruginosa* clone found in CF patients also occurs at a relatively high frequency in rivers, lakes, and sanitary facilities.

Such findings may suggest a threat to human health from in situ bioremediation based on stimulating the growth of indigenous organisms, such as pseudomonads. However, the question of whether precisely the same strains are active in the environment and in causing human infections has been difficult to answer because conventional taxonomic tests can produce ambiguous results.

In their 1996 paper, Westlake and his collaborators discussed their comparison of 27 strains of *P. aeruginosa* isolated from human patients infected by the organism and 15 strains recovered from the gasoline-polluted aquifer that Ridgway's team described in their 1990 report. The clinical specimens came from eight different hospitals in Canada and the United States, about half from CF patients and the rest from a variety of conditions ranging from an infected burn to an ankle ulcer.

All of the strains were subjected to an intensive battery of taxonomic tests. These included not only phage sensitivity, growth temperature range, membrane fatty acid analysis and the presence of plasmids, but also PCR amplification and sequencing of a species-specific 16S-23S rRNA gene internal transcribed spacer region.

The good news to emerge from this scrutiny was that the clinical strains did indeed differ from those recovered from the contaminated aquifer. The organisms from human patients were unable to use gasoline supplied in the gas phase as their sole source of carbon. The environmental isolates did have this ability.

But there was disquieting news, too. The two sets of *P. aeruginosa* strains emerged from their extraordinarily thorough taxonomic examination as indistinguishable from one another. At the very least, this indicated that no genetic marker was available to discriminate between strains with pathogenic potential and those that are truly innocuous. At worst, it suggested that organisms being considered as prime candidates for bioremediation are also opportunistic pathogens.

As a student, I learned that *P. aeruginosa* was a friendly saprophyte, perhaps helping to break down dead matter in rivers and lakes but otherwise doing little except growing and surviving. It occasionally colonized the human gut, and on those even rarer occasions when it entered an open wound, it could cause suppuration with bluish-green pus.

Since those days, the medical significance of *P. aeruginosa* has become much more apparent, especially in causing nosocomial infections. However, only in the last 2 decades have we discovered the extensive roles of this and other pseudomonads in ridding the environment of natural and unnatural toxic substances. Now we have evidence that environmental and clinical isolates of *P. aeruginosa* are taxonomically identical.

There are those who argue that bioremediation should be conducted, not by the somewhat unpredictable strategy of stimulating the activity of the existing flora in a polluted site, but by using inoculants genetically manipulated to have no pathogenic or other harmful potential whatever. Donald Westlake's paper lends strong support to that view.

References

1. **Aragone, M. R., D. M. Maurizi, L. O. Clara, J. L. Navarro Estrada, and A. Ascione.** 1992. *Pseudomonas mendocina*, an environmental bacterium isolated from a patient with human infective endocarditis. *J. Clin. Microbiol.* **30:**1583–1584.

2. **Foght, J. M., D. W. Westlake, W. M. Johnson, and H. F. Ridgway.** 1996. Environmental gasoline-utilizing isolates and clinical isolates of *Pseudomonas aeruginosa* are taxonomically indistinguishable by chemotaxonomic and molecular techniques. *Microbiology* **142:**2333–2340.

3. **Rahme, L. G., E. J. Stevens, S. F. Wolfort, J. Shao, R. G. Tompkins, and F. M. Ausubel.** 1995. Common virulence factors for bacterial pathogenicity in plants and animals. *Science* **268:**1899–1902.

4. **Römling, U., J. Wingender, H. Müller, and B. Tümmler.** 1994. A major *Pseudomonas aeruginosa* clone common to patients and aquatic habitats. *Appl. Environ. Microbiol.* **60:**1734–1738.

64 Yeasts Are Complex . . .

Over 4 decades ago, when I was a research student exploring a minuscule aspect of a minuscule segment of the microbial world, the animalcule I cultured, fragmented, and analyzed each day was *Saccharomyces cerevisiae*. Books, both scholarly and popular, described it as a free-living unicellular organism. Some authors added the adjective "simple" or even "primitive."

Only occasional dissenters questioned the conventional wisdom. They claimed, for example, that microbiologists categorized yeasts as solitary cells simply because this was the form *S. cerevisiae* and its relatives adopted when compelled to proliferate in the artificial environment of the shake flask, fermentor, or chemostat. Maybe yeasts in their own world were more social in habit and/or multicellular in form? I even managed to produce some filamentous versions of *S. cerevisiae* myself, chains of cells that failed to separate when grown in biotin-deficient medium (4).

True, filamentation was mentioned in taxonomic guides, but in practice it was sidelined or overlooked altogether. The prevailing view was that yeasts were "planktonic" creatures, teeming populations of individual cells thriving on the nourishing warmth of laboratory glassware. As microbiology progressed, these were the organisms that spawned numerous advances in biochemistry and other specialties.

Their unicellular habit made them ideal experimental material precisely because one cell did not communicate with another. Especially prized was *Schizosaccharomyces pombe*, which provided major insights into the operation of the eukaryotic cell cycle.

Is the textbook description of yeasts still appropriate, though? Recent years have provided several clues that point in a rather different direction. Consider, for example, the likelihood that apoptosis occurs in these allegedly solitary, elementary organisms.

Apoptosis (often said to be synonymous with programmed cell death, though there are other forms of the latter) is inherently a property of multicellular animals. Distinguished from necrosis because it occurs without an associated inflammatory response, it has important roles during embryogenesis and other processes when cells need to die in a manner useful to the animal as a whole. Apoptosis begins with phosphatidylserine flipping to the outer leaflet of the plasma membrane. The cell shrinks and collapses in an orchestrated way, as the membrane blebs, chromatin condenses, and DNA fragments. Then, neighboring cells quickly engulf the moribund cell.

Although delineation of the *S. cerevisiae* genome has not revealed any genes coding for the type of proteins involved in apoptosis in metazoa, there have been other indications that the process may occur in yeast. Particularly persuasive is the demonstration that inactivation of Cdc13, yeast's telomere-binding protein, triggers apoptotic signals, including the flipping of phosphatidylserine (6).

Flirting with teleology at this point, we confront a puzzle. Through the selective, genetically regulated demise of specific members of a population of cells, the whole purpose of apoptosis is to promote changes that benefit the coordinated community. It comes into play during the embryonic development of human fingers and toes, for example, removing inter-connecting webs. But if yeasts really are free-living unicellular organisms, as we have been assured, what conceivable role could there be for programmed cell death?

Intercellular communication is another issue, raised by Richard Dickinson of Cardiff University in the United Kingdom in a critique of the concept that yeasts are free-living, eukaryotic unicells (3). At the most genuinely elementary level, communication between *S. cerevisiae* cells has been known for some years. There are two haploid mating types, producing corresponding pheromones and having surface receptors for the opposite type. Each pheromone arrests the cell cycle of the opposing type. When this happens, the haploids mate and generate a diploid cell.

Going beyond this primitive sexuality, Dickinson highlighted the recognition of quorum sensing in *Candida albicans*. Chen et al. (2) found that cells of this opportunistic human pathogen, inoculated into fresh medium, showed a lag in growth, which was abolished by tyrosol. "Why should it be an advantage for yeasts cells to grow only when accompanied by many others?" Dickinson asked. "Surely a unicellular organism would 'prefer' to be alone so that it and its progeny consume the available nutrients? This seriously questions whether yeasts live a 'unicellular' existence."

Another phenomenon cited by Dickinson was the behavior of *S. cerevisiae* in the presence of fusel alcohols—isoamyl alcohol and others—end products of amino acid catabolism that accumulate when yeasts are short of nutrients. Responding to their own metabolic products, the organisms begin to grow as strings of elongated cells.

"Filaments contain significantly greater mitochondrial mass and an increased chitin content compared with yeast-form cells," wrote Dickinson. "Unlike the situation in yeast-form cells, the chitin of filaments is not confined to bud scars and the chitin ring between mother and daughter cells is distributed over the majority of the cell surface. Hence, the walls of the filaments have greater strength and rigidity than those of yeast-form cells, which represents an important advantage to the filaments, which can penetrate solid media."

Dickinson also drew attention to research on the production of bicarbonate by *S. cerevisiae* during sporulation (a process limited to normal diploids). This work, he believed, indicates that the yeast is not a free-living unicellular eukaryote but a social organism, whose diploid cells "tell" others to sporulate.

Yeast buffs learned many years ago how to shock their subjects into making spores. The trick is to transfer cells to potassium acetate but to deny them both glucose and sources of nitrogen. When they begin to sporulate, the cells convert some of the acetate into carbon dioxide, lowering the pH. If the medium is buffered, this reduces or prevents spore formation. Also long recognized is the fact that sporulation is optimal only within a range of middling optical densities.

We now know why these two conditions are important. One Japanese group (5) found that *S. cerevisiae* cells growing on acetate and starting to form spores use the bicarbonate produced to signal other cells to do the same thing. Bicarbonate, added to the medium, encourages sporulation at low cell densities. It even induces "sporulation-defective" mutants to generate spores.

Putting all of this evidence together, Dickinson made a compelling case for yeasts to be categorized not as free-living unicellular eukaryotes but as "social, colonial organisms with cell-to-cell communication." Thinking again about my strings of biotin-deficient *S. cerevisiae* cells, I wonder whether filamentous growth was a trait generally bred out of laboratory strains because it was an inconvenient form for experimental studies. Indeed, the "rediscovery" of filamentation in the last decade has been accompanied by recognition that yeasts also thrive as biofilms. There is more, much more, to these animalcules than a solitary, uncommunicative habit.

"The temptation to regard small size, and simplicity of structure, whether real or apparent, as criteria of a primitive condition, has often proved the cause of error and confusion," wrote Kenneth Bisset in *The Cytology and Life-History of Bacteria* (1) in 1950. He reserved particular scorn for impressions of the nature of bacteria derived from staining methods whose deficiencies were only then being recognized. "The distorted vestiges of bacteria which survived the technique of drying and heat-fixation were accepted as truly indicative of the morphology of living organisms," he wrote. The artificiality of these views had "served to widen the gap between bacteriology and other biological sciences, as well as to confuse and retard the advance of bacteriology itself."

Though Bisset was something of a maverick, many of his criticisms proved to be highly apposite. Have we been guilty of a similar, if less severe, calumny as regards the mode of life of yeasts?

References

1. **Bisset, K.** 1950. *The Cytology and Life-History of Bacteria.* E&S Livingstone, Edinburgh, United Kingdom.

2. **Chen, H., M. Fujita, Q. Feng, J. Clardy, and G. R. Fink.** 2004. Tyrosol is a quorum-sensing molecule in *Candida albicans. Proc. Natl. Acad. Sci. USA* **101:**5048–5052.

3. **Dickinson, J. R.** 2005. Are yeasts free-living unicellular eukaryotes? *Lett. Appl. Microbiol.* **41:**445–447.

4. **Dixon, B., and A. H. Rose.** 1964. Observations on the fine structure of *Saccharomyces cerevisiae* as affected by biotin deficiency. *J. Gen. Microbiol.* **35:**411–419.

5. **Ohkuni, K., M. Hayashi, and I. Yamashita.** 1998. Bicarbonate-mediated social communication stimulates meiosis and sporulation of *Saccharomyces cerevisiae. Yeast* **14:**623–631.

6. **Qi, H., T.-K. Li, D. Kuo, A. Nur-E-Kamal, and L. F. Liu.** 2003. Inactivation of Cdc13p triggers *MEC1*-dependent apoptotic signals in yeast. *J. Biol. Chem.* **278:**15136–15141.

65 . . . And Yeasts Are Versatile

The Brussels-based European Commission (EC) has often been criticized for using much more energy in announcing its science funding programs than in publicizing the ensuing results. In many cases, indeed, the EC's only tangible credits for fostering many major advances in research have been the footnotes tucked away at the end of journal papers.

"Yeast as a Cell Factory," a symposium held in Vlaardingen, The Netherlands, in 1999, was a timely departure from this pattern. A shop window for EC-supported bioscience, it encompassed 3 days of presentations, including one specifically arranged to inform members of the European Parliament about research on yeast and in allied fields.

Politicians are notorious, in all countries, for failing to find time to attend events of this sort (even those, in my experience, that have been specially organized for them alone). However, what the members of the European Parliament should have taken from the Vlaardingen event is an appreciation of the unity and mutuality of the life sciences today and the key role played by microbiology in that pattern. One example came from a clutch of insights into yeasts, not only as the organisms of choice for making pharmaceuticals in the future, but also as keys to the understanding of a variety of human disease processes.

Clearly, *Saccharomyces cerevisiae* and other yeasts, which already generate 24 billion liters of the world's alcohol each year, are beginning to make health care products, too. Symposium participants heard, for example, about South Korean work on the harnessing of yeast to make human parathyroid hormone to prevent osteoporosis and about United Kingdom and Lithuanian research using yeast as an expression system for human polyomavirus major capsid protein, a step toward the evolution of a vaccine and a diagnostic reagent.

Eric de Bruin of Wageningen Agricultural University in The Netherlands presented evidence that *Hansenula polymorpha* could assemble human type I collagen fragments. Supplementing Finnish research on the production of type III collagen by *Pichia pastoris*, this indicated that yeasts can now be harnessed commercially to make collagen for surgical and medical use more efficiently and safely than by existing methods.

Surgeons use collagen, a fibrous protein comprising almost a third of all the protein in the body, for suturing and repair, as well as to prevent hemorrhage. In cosmetic surgery, injections of collagen into the skin can also overcome scars and other deformities. At present, the collagen in preparations used for these purposes comes either from human cadavers or from cows. However, the emergence of AIDS, and more recently bovine spongiform encephalopathy in the United Kingdom, heightened concern about the possibility that material taken from such sources might be contaminated with viruses or prions.

The Dutch meeting heard, however, that both *H. polymorpha* and *P. pastoris* had been genetically engineered to synthesize collagen. Not only did they accomplish the task efficiently, but also, such a process was considered extremely unlikely to be contaminated with the organisms that might be present in the cells of humans and other animals.

Another speaker from Wageningen Agricultural University, Jan Verdoes, described the use of *Xanthophyllomyces dendrorhous* to produce carotenoids for medical and other applications. These pigments had for the most part been manufactured by chemical synthesis. On environmental and economic grounds, however, it would be preferable to produce them by natural means. Verdoes believed that yeasts were the perfect answer.

One particular carotenoid, astaxanthin, is responsible for the pink to red coloration of many fishes, crustaceans, and birds. In the aqua-

Animalcules: the Activities, Impacts, and Investigators of Microbes

culture industry, it is added to trout and salmon feed to ensure that they develop this natural pigmentation. The red yeast *X. dendrorhous* also makes astaxanthin, however, though only in relatively small quantities.

Verdoes and his colleagues used carotenoid biosynthetic genes from the bacterium *Erwinia uredovora* to isolate several genes from *X. dendrorhous* coding for enzymes responsible for different stages in the biosynthesis of astaxanthin. Some of these genes were reintroduced, and as a result, the yeast produces considerably more of the pigment than strains not altered in this way. Astaxanthin also functions as an antioxidant. This means that it may in future find uses in the prevention of cancer and degenerative diseases

Johan Thevelein of the University of Leuven and the Flemish Interuniversity Institute of Biotechnology in Belgium reported on the glucose-sensing system for activation of the cyclic AMP pathway in yeast. He and his collaborators found that *S. cerevisiae* cells use a receptor protein with the same structure as that in many mammalian hormone receptors for the detection of glucose. Moreover, the signal from the receptor is transmitted in the same way, through a G protein that activates adenylate cyclase.

The twofold implications of this work were an admirable illustration of the inseparability of pure and applied science and of the coherence of medical and other biological research. On one hand, the Belgian investigators believed they might be able to exploit their discovery to genetically engineer strains of yeast for inclusion in frozen dough that do not (as normally happens) lose much of their capacity for fermentation during storage. They removed from *S. cerevisiae* the receptor protein that senses glucose and thus delayed its loss of freeze resistance.

On the other hand, the discovery of the common pathway of glucose detection between yeast and mammalian cells had health implications, too. It suggested that certain "orphan" receptors on human cells, whose purpose was hitherto unknown, might in fact serve to detect nutrients. Thevelein argued that, since the glucose receptor in yeast activates a signal transduction pathway used in most cells to transmit extracellular information, and by a common mechanism, nutrients might interfere in this way with the major transduction pathways regulating the functioning of cells in the human body.

It seemed possible that this finding could account for some of the odd effects of certain foods and diets in some individuals. Inexplicable on the basis of our understanding of nutrients serving solely as sources of energy and growth substrates, these effects might reflect a previously unknown realm of interaction with the human endocrine system.

Finally, a European meeting on yeast would have been unthinkable without papers on alcoholic beverages. Here, too, major advances were reported, ranging from Belgian progress in speeding cider production by continuous fermentation to German (yes, German) engineering of *S. cerevisiae* to make less (yes, less) ethanol in low-alcohol beers. Sylvie Dequin showed how she and colleagues at the National Institute for Agricultural Research in Montpellier, France, constructed strains of *S. cerevisiae* that produce more glycerol, and less acetic acid, than usual. They believed that yeasts altered in this way might be of considerable value in improving wine quality. Alongside alcohol, glycerol is the main by-product of fermentation, giving both sweetness and fullness to wines. Genetically engineered yeasts should be particularly welcome in regions with a cool climate, to help in making wines with greater body than was achievable in the past.

The key feature of the new strains was overexpression of the gene responsible for glycerol-3-phosphate dehydrogenase, the activity of which determines the amount of glycerol a yeast produces. The new strains thus generated two to three times as much glycerol as the original strains, and slightly less alcohol. A further advantage of the genetically modified yeasts is that fermentation is more vigorous than before, saving time and thus reducing the costs of the process.

66 Resounding Banalities

What is the point of anyone writing, publishing, or reading opening sentences such as these:

Travellers often develop diarrhoea during stays in tropical and subtropical destinations. (9)

Plant oils and extracts have been used for a wide variety of purposes for many thousands of years. (7)

The use of pesticides has become an integral part of modern agricultural systems. (6)

Zoonoses are diseases that can be transmitted from animals to humans and are public health threats worldwide. (8)

The Nobel laureate Peter Medawar once suggested that authors should never launch a scientific paper with a resounding banality. He had in mind grandiose statements such as "Malaria is a mosquito-borne disease which is responsible for considerable annual burdens of both morbidity and mortality among people living in tropical regions of the world."

A moment's reflection tells us that this type of opener serves no purpose whatever. Not only malariologists, not only protozoologists, not only microbiologists, but virtually everyone knows that malaria is transmitted by mosquito bites, that it occurs in the tropics, and that it can make you ill and die. So why waste words? Similarly, why tell a 21st-century audience that farmers use pesticides, that humans have always used plant oils, and that travelers to the tropics get diarrhea?

We all tend to write such sentences, of course. They emerge from the cerebral cortex as soon as we settle down at the screen, clear the throat and try to command our readers' attention. What we need to do, therefore, is to recognize the fatuity of such declamations and use the delete key. If we are seriously unaware of the problem, editors should do the surgery instead.

Resounding banalities are not merely purposeless. They positively impair the quality of scientific writing by replacing what could be far more enticing overtures. Consider the following, which are genuinely interesting:

Fungi are an emerging cause of hospital-acquired infection. (1)

More than seven decades after its discovery, the bactericidal effects of penicillin remain mysterious. (3)

What is systems biology, with its steep learning curve, good for? (2)

Each of these facts and ideas will doubtless have been familiar to some readers, but by no means all. For many others, the sentences contain, respectively, an arresting fact, a major surprise, and a provocative question.

That is the crucial point. Every research paper will appeal to a well-defined core of specialists. However, an attractive title, opening sentence, and first paragraph can, in addition, reach a much wider penumbra of readers. This will include browsers and people in adjacent disciplines.

An author submitting a report to a dedicated journal of antibiosis, for example, does not need to highlight the significance of developments in understanding penicillin action. Some specialists will be obliged to read such a paper (even one that is tediously written) because the topic is so close to their own work.

The introduction to a similar paper appearing in *Trends in Microbiology*, on the other hand, could be designed to tweak the antennae of other potential readers. Many of them, preoccupied by adenoviruses or bioremediation, may wrongly believe that the action of penicillin on cell wall synthesis was sorted out decades ago. They need an opening that builds a bridge, explaining why both the subject and findings are important.

It is because much potentially rewarding reading is not obligatory but serendipitous that introductory sentences and paragraphs are so crucial. Here is Bryn Bridges, launching a commentary on adaptive mutation in *Trends in Microbiology* (4):

Bacteria, like some actors, tend to spend most of their time "resting." They often do not know where their next meal is coming from, and show no obvious sign of activity to the casual observer. But, just as "resting" actors may in fact be very busy, eking out their resources and trying new approaches, so "resting" bacteria call into play a whole new set of metabolic activities that are distinct from those occurring during active growth.

Closer in tone to a popular magazine article than a scholarly paper, this makes a refreshing change from learned journals' standard fare. Although such an approach would not be appropriate or necessary for every communication, it was highly effective here.

Bridges knew that he did not need to arouse the curiosity of bacteriologists already acquainted with so-called directed mutations. He recognized, however, that *Trends in Microbiology* readers also include virologists, mycologists, and others who might, with a little help, find the subject absorbing, so he thought carefully about comparative or metaphorical language to illuminate his subject for a wider audience.

The more heterogeneous that readership, the greater the opportunities to extend helping hands in this way. Here is Richard Cammack, opening a piece in *Nature* (5) on methyl-coenzyme M reductase: "Whenever biomass is degraded, and oxygen runs out, methanogenic bacteria appear and make methane in profuse quantities. For us it is a potential fuel. For the microbes, it is the ultimate waste product. The structure of the enzyme that produces it. . . ."

Here a thought-provoking contrast between "fuel" and "waste product" has replaced the pedestrian or highly technical introduction one might expect for such a topic. Both *Nature* and *Science* use devices of this sort very effectively in their news sections to introduce concepts, theories, and even techniques to a larger audience than would otherwise encounter them. Their approach could be adopted far more widely and with great benefit across the entire range of journals.

Microbiologists are, of course, no more blameworthy for resounding banalities and impenetrable writing generally than people in psychology or plate tectonics. On the other hand, there appears to be little recognition of the problem.

Why, within microbiology, are there no initiatives such as that of Chicago's Fermilab, whose particle physicists now have to produce "plain English" Web versions of their papers to attract scientists from other fields? Why do general microbiology journals not follow *The Lancet* in advising authors to try out their first drafts on researchers in other fields before submission?

One final suggestion. Those who simply cannot suppress the urge to brandish assertions of the unquestionable, the tautological, the fully familiar, and the utterly boring should finish, not begin, their papers in this way. Dullness does not matter so much down there.

References

1. **Arvanitidou, M., K. Kanellou, V. Katsouyannopoulos, and A. Tsakris.** 1999. Occurrence and densities of fungi from northern Greek coastal bathing waters and their relation with faecal pollution indicators. *Lett. Appl. Microbiol.* **29:**81–84.

2. **Baskin, C. R., and M. G. Katze.** 2008. Systems biology could help us understand, protect against pandemics. *Microbe* **3:**227–233.

3. **Bayles, K. W.** 2000. The bactericidal action of penicillin: new clues to an unsolved mystery. *Trends Microbiol.* **8:**274–278

4. **Bridges, B. A.** 1995. Sexual potency and adaptive mutation in bacteria. *Trends Microbiol.* **3:**291–292.

5. **Cammack, R.** 1997. The enzyme at the end of the food chain. *Nature* **390:**443–444.

6. **Chen, W., and A. Mulchandani.** 1998. The use of live biocatalysts for pesticide detoxification. *Trends Biotechnol.* **16:**71–76.

7. **Hammer, K. A., C. F. Carson, and T. V. Riley.** 1999. Antimicrobial activity of essential oils and other plant extracts. *J. Appl. Microbiol.* **86:**985–990.

8. **Punda-Polic, V., B. Luksic, and V. Capkun.** 2008. Epidemiological features of Mediterranean spotted fever murine typhus, and Q fever in Split-Dalmatia County (Croatia), 1982–2002. *Epidemiol. Infect.* **136:**972–979.

9. **Von Sonnenburg, F., N. Tornieporth, P. Waiyaki, B. Lowe, L. F. Peruski, Jr., H. L. DuPont, J. J. Mathewson, and R. Steffen.** 2000. Risk and aetiology of diarrhoea at various tourist destinations. *Lancet* **356:**133–134.

67 Microbiology
Present and Future

Microbiologists could be forgiven for feeling schizophrenic about their profession and hesitant about asserting how it is likely to develop in this millennium. On one hand, they sense that microbiology is at least as important to the future of the world as any other branch of science, and more so than most. The same comparison highlights the extraordinary vitality and intellectual coherence of present-day microbiology. On the other hand, very public signs have appeared in recent years to indicate that the subject has been absorbed or eclipsed by new, more meaningful disciplines.

In at least one conspicuous sense, it scarcely even exists. Forty years ago, virtually every university that embraced science boasted a department of microbiology. Some dated back to the very origin of that word. Walk around the same campuses today, however, and you will find most of those departments gone. Many have metamorphosed into units and centers of biotechnology or cell biology. Elsewhere the staff have been corralled together with zoologists and botanists and reappeared under banners such as life sciences. In some places, it is now impossible to find a door or notice board sporting the traditional label "microbiology."

Such changes could be simply local organizational shifts sanctioned by committees in response to what they imagine to be fashionable thinking. However, the fact that the same trend is apparent in very

Animalcules: the Activities, Impacts, and Investigators of Microbes

different countries throughout the world implies a good deal more. It seems to signify an underlying reality about the nature of the discipline. The suggestion is that microbiology as a category is dead.

Nonsense. These recent developments establish nothing of the sort. They merely reflect the capricious restlessness that afflicts all branches of human knowledge from time to time. In another 20 years time (I forecast), the changes will have been reversed. Microbiology will be the name on the door once again.

There are two principal reasons for making this assertion. They also provide grounds for arguing that the subject ought to be proclaiming its nature and worth more vigorously, as against other branches of science, than it is at the present time.

First, and despite those external appearances, microbiology remains a peculiarly distinctive craft, robustly free of the self-doubts to be found elsewhere in the scientific firmament. Think of genetics, for example, which even its own practitioners now look upon with mixed feelings. Writing in *Heredity*, the distinguished British geneticist John Fincham had this to say about his subject in 1993 (1):

Whereas in 1945, genetics was a minority interest among biologists, viewed by most as specialised, unintelligible or just irrelevant, it has now penetrated into every corner of biological science. Most biologists would now recognise it as being at the root of everything. This is a source of great satisfaction, but it also creates something of an identity crisis for geneticists. What is genetics? How do we distinguish ourselves from molecular biologists or biochemists, when half the papers in any leading molecular biological or biochemical journal are about the identification of gene products and controls of gene expression?

In concluding his paper, Fincham asked, "Do we any longer have an identifiable subject?" His answer was not a substantive one, founded upon the essential characteristics of his craft. It was purely operational. "Perhaps we should stop worrying," he said, "and adopt a pragmatic definition of a geneticist, perhaps as a person who thinks it worthwhile to subscribe to a Genetical Society."

Now that molecular genetics has influenced the entire range of life sciences, whose exponents are busily locating, cloning, and sequencing genes of interest, it is easy to understand this dilemma of identity. Biochemists have a similar though lesser problem in defining what they are about. So do many other disciplines, including chemistry, which is challenged more acutely than ever to distinguish itself from physics.

The difficulties have particular force in the domain of subatomic physics. Here, the methodology and results can often be expressed only vicariously, in terms of mathematics, while the phenomena under investigation are increasingly artificial rather than "real." Even the precise nature of the work is at issue. Disagreements are commonplace between the theoreticians who plan experiments and the machine crews who carry them out in their particle accelerators.

Thus, while the techniques of microbiology have changed considerably over the decades, as have those of genetics, biochemistry, and particle physics, its mission has endured with far greater clarity than those of other domains of science. Microbiologists study microscopic life. They seek to understand how bacteria, viruses, and other microorganisms behave and how they affect the world, especially our world, for good or ill. Moreover, whether in the realms of biofilms or probiotics, ruminant digestion or bioremediation, biological control or gene transfer in nature, microbial ecology or mixed and opportunistic infections, they are more concerned than ever with microbial cooperation and antagonism in communities and consortia. What could be clearer?

The second reason for highlighting the identity and value of microbiology is its practical significance. All terrestrial life rests upon microbiological activity, which we are also just beginning to learn to harness for human and environmental benefits, too. At the same time, microbes have unrivaled power over other species: they possess the pathogenic and genetic potential to wipe out the human race. To say the very least, this gives their investigation a very different order of importance than most other sciences.

Consider the global relevance of ergonomics or rheumatology, acoustics or aeronautics, pharmacy or pharmacology. In contrast to these valuable though limited pursuits, microbiology is the core discipline required for us to meet two major challenges we now face.

The first is the threat, urgently announced in many papers and reports in recent years, of new, resurgent, and as yet unanticipated human infections. The second is the need to use microorganisms on a far greater

Animalcules: the Activities, Impacts, and Investigators of Microbes

scale than in the past to evolve clean industrial technologies. The goal is to overcome the pollution and resource problems created by the industrial revolutions of the 18th, 19th, and 20th centuries.

Couple these two agendas together and we have a program arguably far more pressing than that of any other branch of science at the present time.

When he delivered his Rede Lecture in Cambridge in 1959, the English novelist C. P. Snow described what he termed the two cultures and thereby launched a debate that has continued to this day. Snow claimed that there was a "gulf of mutual incomprehension" between scientists and "literary intellectuals," which often hardened into mutual hostility, especially among the young. The section of his talk that particularly angered some opponents was that in which Snow touched upon the singular nature of the work of exploration upon which scientists are embarked. "If I were to risk a piece of shorthand," he said, "I should say that naturally they had the future in their bones."

Many of his literary critics felt that Snow was being unpardonably arrogant in appropriating to scientists intuitive feelings of their historic significance. Why should a Darwin or an Einstein be considered to have "the future in their bones" any more than a Shakespeare or a Tolstoy?

It is, of course, foolish to compare the contributions of great writers and scientists, as though they were rivals whose achievements can be measured on a single scale calibrated in units of eternal merit. Snow was certainly not advocating this. Nevertheless, he made a valid point. Scientists can be legitimately proud to take part, in however small a way, in an enterprise that has already vastly improved the human condition and will continue to do so in future, while at the same time constantly extending our rational understanding of the universe.

By these criteria, I believe that microbiology ranks as a preeminent science of both this century and the next. By the same token, microbiologists do indeed have the future in their bones. I hope this will not seem like unpardonable arrogance.

Reference

1. **Fincham, J. R. S.** 1993. Genetics in the United Kingdom—the last half-century. *Heredity* **71:**111–118.

68 Looking Back

Unfortunately, I have lost my "I've been cited" T-shirt, a nice gift from the Institute for Scientific Information in Philadelphia. The garment would have been useful for a little boastful celebration a few years ago, for there, in the *Journal of Applied Bacteriology* (3), were two citations by microbiologists at the University of Novi Sad in Serbia of papers on biotin metabolism in yeast that Tony Rose and I published over 40 years previously in the then *Journal of General Microbiology* (1, 2).

For readers accustomed to frequent shoals of citations, my excitement over a mere pair will appear excessive. Nevertheless, I was delighted with these reminders of the days when I was trying to set my mouse of knowledge alongside the mountain of science. Having been a grey-suited writer on such matters for many years, rather than a white-coated practitioner, I did not expect to see my name in research journals anymore. However, the citations did set me thinking.

First, it was interesting, in an age of PCR, DNA sequencing, and diffusion gradient chambers, to find authors building on work done with relatively primitive techniques. For example, Rose and I monitored the growth of our *Saccharomyces cerevisiae* using a Hilger Spekker absorptiometer to measure the turbidity of yeast suspensions in Samco tubes in a wooden holder; yes, Samco tubes in a wooden holder. There were

Animalcules: the Activities, Impacts, and Investigators of Microbes

other crudities, and other bits of history, in the department at the University of Newcastle upon Tyne where we worked. The medical microbiologists, for example, were achieving excellent results with wooden incubators, built like rabbit hutches, that could have come down from Louis Pasteur himself.

Secondly, and more importantly, our understanding of biotin had been completely transformed over the intervening decades. That transformation came from the combined insights and techniques of microbiology and other disciplines. The new understanding ranges far and wide, from the function of biotin as the prosthetic group of several carboxylases to its deficiency as the cause of inherited disorders in babies and possibly of antibiotic-associated diarrhea, too.

In the early 1960s, medical textbooks were able to dismiss biotin, as a member of the B complex of vitamins (though also known as vitamin H), in one short sentence. Biotin deficiency was virtually unknown, because the vitamin is ubiquitous in foods, because it is synthesized by bacteria in the intestine, and because it is required in vanishingly tiny quantities anyway. One of very few spontaneous clinical cases ever recorded was that of an eccentric hermit who consumed large quantities of egg white. He would have been fine if he had taken the yolk (which is rich in biotin) as well. But avidin, a protein in the egg white, bound all of the biotin in his otherwise sparse diet. As a result, he became biotin deficient and developed alopecia.

Against this medical background, Tony Rose and I were confident that our work had no health or indeed other practical implications whatever. We simply wanted to know how the biochemistry and ultrastructure of *S. cerevisiae* changed when we deprived the organism of biotin. I saw from our successors in Serbia that these things remained of interest, though they used continuous cultivation rather than the vast numbers of conical flasks we worked with in Newcastle.

Today, biotin deficiency in humans is recognized as a significant clinical problem. The first advances along that road came in the late 1960s and early 1970s, when several research teams independently discovered that the vitamin plays a role in certain carboxylases. The reactions catalyzed by these enzymes were the carboxylation of acetyl-coenzyme A (CoA), which is important in fatty acid synthesis; of propionyl-CoA, important in proprionate oxidation; of 3-methylcrotonyl-CoA in leucine oxidation; and of pyruvate in gluconeogenesis. Studies with both mammalian cells and microorganisms contributed to the emerging picture of biotin as the prosthetic group of these enzymes.

The next step followed the introduction of a new piece of equipment, the gas chromatographic spectrophotometer, into the study of metabolic disorders during the 1970s. One of the inborn conditions recognized for the first time by the use of this instrument was termed 3-methylcrotonyl-CoA carboxylase deficiency. The first infant diagnosed with this disorder had an extensive skin rash and chronic neurological abnormalities but improved dramatically when given biotin orally. He was subsequently found to have low activities of several biotin-dependent carboxylases. Other cases of "multiple carboxylase deficiency" then came to light.

Gradually, it became clear that the syndrome occurs in two different forms. Neonatal multiple carboxylase deficiency, characterized by vomiting, lethargy, and hypotonia, is attributable to a deficiency of carboxylase synthetase activity. The juvenile-onset version, which appears around 3 months, causes symptoms such as rash and alopecia and stems from a defect in the intestinal or cellular transport of biotin.

A rather different scenario came to light in 1981, when clinicians in San Francisco reported biotin deficiency in a child with short-gut syndrome. Much of her small and large bowel had had to be removed owing to disease, and she was being fed intravenously. Symptoms of biotin deficiency appeared but abated rapidly when she was given biotin. It seems that antibiotic therapy suppressed the bacteria in the patient's remaining intestine that would otherwise have provided her adequate quantities of the vitamin.

The techniques of microbiology have also played a fundamental role in these studies. Most investigators used a microbiological assay to estimate the concentrations of biotin in plasma and other fluids. The two organisms most favored for this purpose were *Lactobacillus plantarum* and *Ochromonas danica*, whose growth provides an extremely sensitive measure of the level of the vitamin.

Most recently, of course, the astonishing affinity between biotin and either avidin or streptavidin (produced by the soil bacterium *Streptomyces avidinii*) has formed the basis for a number of techniques used in microbiology. Very few other covalent interactions in nature come close to the dissociation constant of 10 to 15 M for interactions between the vitamin and the four binding sites on the avidin or streptavidin molecule. Biotin may therefore be used as a covalent label for (macro)molecules, which can be detected with labeled avidin or streptavidin. First exploited in immunochemical tissue staining, this remarkable affinity is now used to enhance signal detection in immunoassays and DNA probe diagnostics.

I am glad to have had a toehold in the literature of biotin. The story has certainly changed dramatically since my busy days with the Samco tubes, and one development a few years back was especially intriguing. Writing in the *Journal of Medical Microbiology* (4), K. Yamakawa and colleagues, of Kanazawa University, Kanazawa, Japan, reported that *Clostridium difficile* in suboptimal levels of biotin grew more slowly but produced greatly increased quantities of its two toxins. This could help to explain why the organism causes antibiotic-associated diarrhea: antibiotic therapy impairs biotin production by the intestinal flora.

Meanwhile, I still wonder about one aspect of my work with Tony Rose that I have not since encountered in the literature. When we grew our yeast in suboptimal concentrations of biotin, we found that it always produced a characteristic pink pigment. I would love to know if anyone has ever extracted and identified this product. Tony called it nipple pink.

References

1. **Dixon, B., and A. H. Rose.** 1963. Effect of biotin deficiency on feedback mechanisms controlling the synthesis of ornithine carbamoyltransferase by *Saccharomyces cerevisiae. J. Gen. Microbiol.* **31**:32.

2. **Dixon, B., and A. H. Rose.** 1964. On the synthesis of ornithine carbamoyltransferase in biotin-deficient *Saccharomyces cerevisiae. J. Gen. Microbiol.* **34**:229–240.

3. **Pejin, D., and R. Razmovski.** 1996. Continuous cultivation of *Saccharomyces cerevisiae* at different biotin concentrations in nutrient media. *J. Appl. Bacteriol.* **80**:53–55.

4. **Yamakawa, K., T. Karasawa, S. Ikoma, and S. Nakamura.** 1996. Enhancement of *Clostridium difficile* toxin production in biotin-limited conditions. *J. Med. Microbiol.* **44**:111–114.

69 A Global Challenge

The main refuge for waterfowl in Europe, one of a series created across the continent to provide sanctuary for birds flying on their long migrations, is Doñana National Park in southwestern Spain. It receives continual expert attention from both ornithologists and biologists concerned with other aspects of conservation. Also designated a UNESCO World Heritage Site, Doñana is home to representatives of more than 70% of all European bird species.

Paradoxically, it was here that thousands of waterfowl died recently in an incident considerably worse than the mass mortalities that occur from time to time along migratory routes. It began when blue-green scum, composed of dense patches of cyanobacteria, appeared on the Los Ansares lagoon and elsewhere in the park. Three days later, observers found thousands of fish floating on the lagoon and hundreds of dead herbivorous waterfowl nearby. Then, other birds died, presumably as a result of eating the dead fish. Within 2 weeks, a total of at least 6,000 birds, including endangered species, such as the white-headed duck and marbled teal, had succumbed.

When investigators from the Universidad Complutense de Madrid in Madrid, Spain, sampled the Los Ansares lagoon, they found that the main and extremely abundant phytoplankton species was a toxin-producing cyanobacterium, *Microcystis aeruginosa*. The water contained high concentrations of cyanotoxins, in particular, a large quantity of microcystin. Clinical signs shown by the dying birds, together with post-mortem findings, were all consistent with cyanotoxicosis.

"The severe cascade of deaths in the Doñana National Park can be explained by the role of cyanotoxins in the food web," wrote the investigators (5). "First, cyanotoxins affect bird species that consume the cyanobacterial scum. At the same time, cyanotoxins accumulate in zooplankton and aquatic invertebrates; hence the cyanotoxins affect fish that feed on plankton. Finally, piscivorous birds consume cyanotoxins in the contaminated fish." The conclusion was that the levels of microcystin detected in the livers of the dead birds and fish were sufficient to explain the mass mortality.

This was not the first such incident to occur in a wildfowl reserve like Doñana. Indeed, the gathering of huge numbers of birds at a specially created and protected wetland site is likely to heighten their vulnerability to disease. Still, researchers are beginning to wonder whether climatic change may be one factor responsible for events of this sort.

Recent years have certainly seen growing anxiety about cyanobacterial threats to aquatic ecosystems, ranging from Lake Taihu in China and Lake Victoria in Africa to Lake Erie in North America and the Baltic Sea in Europe. In addition to the production of toxins by some cyanobacteria, which can not only harm wildlife but also cause liver, neurological, and other diseases in humans, algal blooms kill fish and invertebrates by depleting oxygen and by increasing the turbidity of water.

Rising temperatures favor cyanobacteria in several ways, as Hans Paerl and Jef Huisman have pointed out (7). They mostly thrive better at higher temperatures (often above 25°C) than other phytoplankton, such as diatoms and green algae. Warming of surface waters also strengthens the vertical stratification of lakes, reducing vertical mixing.

"Furthermore, global warming causes lakes to stratify earlier in Spring and destratify later in Autumn, which lengthens optimal growth periods. Many cyanobacteria exploit these stratified conditions by forming intracellular gas vesicles, which make the cells buoyant," Paerl and Huisman wrote. "Buoyant cyanobacteria float upwards when mixing is weak and accumulate in dense surface blooms. These surface blooms shade underlying non-buoyant phytoplankton, thus suppressing their opponents through competition for light."

There are even examples (as in the Baltic Sea) of cyanobacterial blooms raising water temperature directly, by absorbing light very intensively. All of their deleterious activities may be enhanced by alterations in the hydrological cycle, such as more intense precipitation, caused by global warming.

The scale of the dangers of cyanobacteria for human health is illustrated by the situation in Taihu, China's third largest lake. There, a vast bloom appeared at the beginning of June 2007 and within a year had necessitated the removal of more than 6,000 tonnes of algal sludge. As with many other cyanobacterial blooms, the underlying cause was eutrophication, attributed to an accumulation of nutrient-rich sewage and runoff from agricultural land, but the trigger was unusually hot, dry conditions in the area. Normally, the lake not only serves as China's most important fishery, but also provides drinking water for over 2 million people.

Evidence that global warming is affecting cyanobacterial populations comes from the substantial extension of their geographical ranges. One example is *Cylindrospermopsis raciborskii*, which was responsible for a mysterious outbreak of severe hepatitis-like illness in Palm Island, Australia (1). Originally a tropical and subtropical species, it has moved into higher and higher latitudes and is now widespread in the lakes of northern Germany.

All of these developments point to a new agenda and a new challenge for microbiologists in seeking to understand and, where possible, ameliorate some of the consequences of global warming. Another major component of that agenda is, of course, the direct impact of raised average temperature on human pathogens and their vectors. Whereas just a decade ago, discussion was largely restricted to speculations founded on computer modeling, there are now more tangible grounds for concern.

In its *Fourth Assessment Report on Climatic Change Impacts, Adaptation and Vulnerability* (4), published in 2007, the Intergovernmental Panel on Climate Change concluded that dengue fever was likely to become more common as a result of global warming. The panel also mentioned malaria, which has attracted the most attention in this context, but highlighted dengue, a more urban disease, since climatic change is likely to play an even more important role in its spatial and temporal distribution. One problem is that, while rises in temperature or rainfall can promote the disease, so can drought, because household water storage may provide more breeding sites for mosquitoes.

Writing in *The Lancet* (6), Anthony McMichael and colleagues have argued that, while no one study is conclusive, several reports have already indicated effects of global warming on some infections. For example, in association with alterations in climate, the geographical range of ticks that transmit Lyme borreliosis and viral encephalitis has extended northward in Sweden and increased in altitude in the Czech Republic. Also, changes in the intensity (amplitude) of the El Niño cycle since 1975, and more recently its frequency, have accompanied the strengthening of the relationship between the cycle and cholera outbreaks in Bangladesh.

Similar arguments apply to plant diseases. Neal Evans and coworkers, of Rothamsted Research, Rothamsted, United Kingdom, have demonstrated that global warming will increase both the range and severity of phoma stem canker (2). This could mean a corresponding rise in the worldwide losses of $900 million already caused by epidemics of this disease in oil seed rape and other brassicas.

In the media and popular books and articles, microbiology tends to be scarcely mentioned, alongside climatology, ecology, meteorology, and computer modeling, as a source of key questions, and answers, about global warming and its practical repercussions. Given the multifarious roles of microorganisms as drivers of some of the largest-scale phenomena on the planet, from photosynthesis and the cycling of nitrogen and other elements to pandemics of infectious disease (3), it could soon prove to be one of the most important. There is much microbiology yet to do.

References

1. **Carmichael, W. W.** 2001. Health effects of toxin-producing cyanobacteria: "the CyanoHABs." *Hum. Ecol. Risk Assess.* **7:**1393–1407.

2. **Evans, N., A. Baierl, M. A. Semenov, P. Gladders, and B. D. L. Fitt.** 2007. Range and severity of a plant disease increased by global warming. *J. R. Soc. Interface.* doi:10.1098/rsif.2007.1136.

3. **Falkowski, P. G., T. Fenchel, and E. F. Delong.** 2008. The microbial engines that drive Earth's biogeochemical cycles. *Science* **320:**1034–1039.

4. **Intergovernmental Panel on Climate Change.** 2007. *Fourth Assessment Report on Climatic Change Impacts, Adaptation and Vulnerability.* Intergovernmental Panel on Climate Change, Geneva, Switzerland.

5. **Lopez-Rodas, V., E. Maneiro, M. P. Lanzarot, N. Perdigones, and E. Costas.** 2008. Mass wildlife mortality due to cyanobacteria in the Doñana National Park, Spain. *Vet. Rec.* **162:**317–318.

6. **McMichael, A. J., R. E. Woodruff, and S. Hales.** 2006. Climate change and human health: present and future risks. *Lancet* **367:**859–869.

7. **Paerl, H. W., and J. Huisman.** 2008. Climate: blooms like it hot. *Science* **320:**57–58.

Index

Antibiotic resistance
  agents causing, 87–90
  in bird microorganisms, 4–6
  causes of, 179–180
  Denmark policy on, 118–121
  in kitchen biocides, 105–109
  in natural disasters, 92–93
  plant-human transfer of, 279–281
Ants, fungus gardens of, 251–253
Apoptosis, in *Saccharomyces cerevisiae*, 301
Aquaculture, fish pathogens in, 270–274
Arcobacter, in seawater, 86
Art work, deterioration and restoration of, 13–16
Ascomycetes, lichen association with, 18–20
Asparaginase, for acrylamide reduction, 55
Astaxanthin production, 45–46, 306–307
Astroviruses, in mink farms, 286
Attine ants, fungus gardens of, 251–253
Austin, Brian, on fish pathogens, 271–272
Autographa californica, genetically modified, 134–135
Avian influenza, 103, 267–268
Avoparcin, in pig fodder, 118

B
Bacillus subtilis, for fish infections, 271–272
Bacteriophages
  component transfer of, 278
  discovery of, 228–232
  periodic variations in, 80–81
Baculoviruses, genetically modified, 134–136
Baird, Bill, on star jelly, 32–33
Baldry, Peter, on pathogen extermination, 215
Banalities, in writing, 309–313
Barnard, Christiaan, heart transplantation by, 152–153
Bastian, Henry, on spores, 222–223
The Battle Against Bacteria (Baldry), 215
Bdellovibrio, in biofilms, 76

Beck, William S., on Fibiger, 224
Beer production, 60, 308
Behring, Emil von, 225
Beijerinck, Martinus, 255, 259
Belt, Thomas, on ant fungus farming, 251
Benediktsdóttir, Eva, on fish pathogens, 273
Bergey, David H., on *Streptococcus hemolyticus,* 257
Biocides, in kitchen, 105–109
Biodiversity, 22–25, 69–73
Biofilms
  biodiversity in, 69–73
  examples of, 74–78
Bioremediation
  of art work deterioration, 13–16
  of petroleum spills, 296–299
  public attitude toward, 189–193
Bioterrorism, P4 high-containment laboratories and, 288–289
Biotin, 318–321
Birds
  as disease vectors, 3–7
  influenza in, 103, 267–268
  large-scale die-off of, 322–323
  psittacosis in, 143–144
  Usutu virus in, 284–285
Bishop, David, on scorpion toxin in pesticides, 133–137
Bisset, Kenneth, on bacterial distortion in fixation, 291, 303
Bitting, Katherine G., on organisms in home canning, 257–258
Black-headed gulls, as disease vectors, 4–7
Blaser, Martin, on *Helicobacter pylori,* 124
Bock, Eberhard, on air pollution protection, 15–16
Books, fungal infections of, 43–44
Borovsky, P. F., 248
Borrelia burgdorferi, 138–141
Botryosphaeria rhodina, in biofilms, 77
Botulinum toxin, 34–38

Bovine spongiform encephalopathy, 185–187, 279–280

Boyd, William, *Pathology for the Physician*, 41–42

Bray, John, on *Escherichia coli*, 49

Breed, Margaret S., on taxonomy, 256–257

Breed, Robert S., on taxonomy, 256–257

Breitbart, Mya, on viruses, 80

Bridges, Bryn, on adaptive mutation, 311

Brown, Michael, on microorganism evolution, 122–125

Bruce-Chwatt, Leonard, on Hoare, 248

Brucella abortus, discovery of, 127

Buellia frigida, sexuality of, 17

Buildings, air pollution action on, 15–16

Burgdorfer, Willy, in Lyme disease organism discovery, 140

Burke, Derek, on interferon, 235

Burkholderia cepacia, plant-human crossovers of, 279–281

Burnet, Macfarlane
on communicable diseases, 99
on Isaacs, 233–234
on myxomatosis, 176–177

C

Caliciviruses
outbreaks of, 285
in walruses, 286

Camilli, Andrew, on virulence, 127–128

Cammack, Richard, 311–312

Campylobacter
birds disseminating, 4–6
in seawater, 86

Camus, Albert, *La Peste,* 39–40

Cancer
caused by nematodes, 224–227
vaccines for, 181–182

Candida, in acetaldehyde generation, 168

Candida albicans, quorum sensing in, 302

Cano, Raul, 260

Cantell, Kari, on interferon, 235

Carcinogens, acetaldehyde as, 166–169

Cardboard, cellulolysis of, 26–29

Carlsson, Victoria, 257

Carotenoid production, 306–307

Carr, Noel, 260

Carroll, James, 211

Carter, K. Codell, on Koch's postulates, 206

Castellani, Aldo, 258

Cattle
botulism in, 34–38
Escherichia coli in, 50
foot-and-mouth disease in, 96–99, 100–101
mad cow disease in, 185–187, 279–280
rumen organisms in, 54–55

Cecina (beef product), 44–45

Cellini, Luigina, on *Helicobacter pylori* in seawater, 84–86

Cellulolysis, 26–29

Cellulose, modified, 57–58

Cellvibrio, in cellulolysis, 27

Cervical cancer, vaccines for, 181

Cha, Hyung Joon, on adhesives, 56

Chain, Ernst, 245

Chanock, Robert, 260

Characklis, William, on biofilms, 70

Chase, Martha, on bacteriophages, 229

Cherwell, Lord, security pass used by, 288

Chiang Kai-shek, Churchill meeting with, 242

Chlamydophila psittaci, in veterinary teaching hospital, 143–144

Cholera, *see Vibrio cholerae*

Churchill, Winston
security pass used by, 288
sulfa drugs for, 242

Cigar manufacture, fermentation in, 53–54

Citation analysis, 194–198

Citrobacter rodentium, phages of, 81

Citrus juices, bitter substances in, 45

Cladonia, lichen association with, 19

Clark, Paul F., on Noguchi, 212

Cloned animals, 164

Clostridium, in cellulolysis, 28

Clostridium botulinum toxins, 34–38

Clostridium proteoclasticum, in rumen, 54–55

Clostridium tetani, extermination of, 116–117

Clothing, laundering, microorganisms surviving after, 290

Cobo, Fernando, on stem cell contamination, 144–145

Cognitive dissonance, in infectious disease treatment, 62–64

Cohen, Barnett, 258

Cohen, Sheldon, on psychoneuroimmunology, 64

Cohn, Ferdinand Julius, 220–223

"Cold-water disease," of fish, 273

Cole, Eugene, on home biocides, 107–108

Colitis, hemorrhagic, 48

Collagen synthesis, using yeasts, 306

Colwell, Rita, 260

Common cold, psychoneurology of, 61–64

Concha, Angel, on stem cell contamination, 144–145

Conn, Harold J., 256

Conn, Herbert W., 254–256

The Conquest of Epidemic Disease (Winslow), 215

Contact lens storage, biofilms in, 70–71

Copper, antibiotic resistance due to, 87–90

Corked wine, 43–47

Corneal transplantation, 152–153

Coronary artery bypass surgery, 83–84

Corynebacterium, in acetaldehyde generation, 168

Corynebacterium ammoniagenes, in cigar manufacture, 54

Costerton, William, on biofilms, 70–71

Cowpats, research on, 266–268

Cows, *see* Cattle

Craig, Wallace, on psychoneuroimmunology, 61–64

Cremers, Hayo Canter, on escape of genetically engineered organisms, 161–162

Creutzfeldt-Jakob disease, variant, 185–187, 279–280

Crittenden, Peter, on lichen sexuality, 18–20

Cummings, Stephen, on cellulolysis, 28–29

Curtiss, Roy, 260

Cyanobacteria
toxic products of, 322–324
virus symbiosis with, 23

Cyanophages, 81–82

Cyclidium glaucoma, habitats of, 24

Cylindrospermopsis raciborskii, in lake water, 324

Cytophaga, in cellulolysis, 27

D

Dagley, Stanley, 260

Daims, Holger, on wastewater treatment, 59

Dantas, Gautam, on disasters, 93–94

Dark, Alvin, 212

Darwin, Charles
Hoyle challenge to, 250–253
on locust swarms, 8
on rabbit evolution, 177

Davies, David, on biofilms, 70–72

de Bruin, Eric, on collagen synthesis using yeasts, 306

De Graaf, Regnier, Leeuwenhoek association with, 202

de Kruif, Paul, *Microbe Hunters,* 41, 220

Debaryomyces hansenii, in cigar manufacture, 54

Defoe, Daniel, *A Journal of the Plague Year,* 40

Delbruck, Max, on phages, 228

Delisea pulchra, biofilms on, 76–77

Delwiche, Eugene A., 260

Demain, Arnold, 260

Kennaway, Sir Ernest, on cancer causes, 226

Kermack, W. O., on epidemiology, 100

Keunen, J. G., 260

Kilbourne, Edwin, 260

Kissen, David, on psychoneuroimmunology, 63–64

Kitchen biocides, 105–109

Klebs, Edwin, 206

Klebsiella pneumoniae
    in locust swarming, 10–11
    resistance in, 88

Koch, Robert
    on cholera, 239–240
    Cohn encouragement of, 223
    Fibiger association with, 225
    Nobel Prize of, 239
    pioneering work of, 220
    postulates of, 206–209

Koide, S. S., on Noguchi, 213

Koprowski, Hilary, 260

Kundsin, Ruth, 260

L

La Peste (Camus), 39–40

Laboratories, high-containment P4, 288–289

Lactic acid bacteria, for fish infections, 271

Lactobacillus plantarum, in biotin assay, 320

Lainson, Ralph, trypanosome discovery by, 246

Lambert, Cary, on biofilms, 76

Lampen, Joe, 260

Landfills, paper degradation in, 28–29

Lanois, Anne, on *Steinernematidae*, 287

Lappin-Scott, Hilary, on biofilms, 71

Laundry, microorganisms surviving after, 290

Laybourn-Parry, Johanna, on viruses, 82

Leaning, Jennifer, on disasters, 94

Leask, Ronnie, on star jelly, 32–33

Leeuwenhoek, Antony van, 201–205

Legionella pneumophila, in protozoa, 123–124, 282

Leishmania, viruses in, 286–287

Lennette, Edwin H., 260

Leptospira noguchi, 213

Leucoagaricus gongylophorus, in ant gardens, 251–252

Levy, Stuart, on home biocides, 106

Lichens, sexuality of, 17–21

Lilly Award of 1950, 259

Limonoids, bitterness of, 45

Lindegren, Carl, 258

Lindenmann, Jean
    citation of, 194
    on interferon, 233–237

Lister, Joseph, challenges to, 238

Listeria monocytogenes
    alcohol interactions with, 55–56
    in cold environment, 290
    in washed vegetables, 145

Literature, infectious disease descriptions in, 39–42

Livestock, antibiotics for, 118–121

Lizards, *Mycobacterium marinum* in, 142–143

Locust swarming, microorganism involvement in, 8–12

Lyme disease, public interest in, 138–141

M

Maalin, Ali Maow, as last smallpox victim, 114

Mackay, Ian, on human metapneumovirus, 285

Mad cow disease, 185–187, 279–280

Malachite green, for fish infections, 271

Manefield, Mike, on quorum sensing, 277

Manure, research on, 266–268

mar operon, for antibiotic resistance, 87–88

Marine and Freshwater Microbiology Biodiversity program, 22–25

Marshall, Barry, on *Helicobacter pylori*, 83–84

Matin, A. C., 260
Mayr, Ernst, on species definition, 222
McCollum, Elmer, on pasteurization, 151
McLaughlin-Borlace, Louise, on biofilms, 71
McMichael, Anthony, on global warming, 325
McSweegan, Edward, on Lyme disease vaccine, 141
Measles
    in disasters, 92
    vaccination for, 102–103, 171–174
Meat
    botulinum toxin in, 37
    foot-and-mouth disease and, 96–99
    mad cow disease and, 185–187
    Salmonella enterica serovar Typhi in, 159–160
Medawar, Peter, on scientific writing, 309
Mediterranean Sea, *Helicobacter pylori* in, 84–86
Meetings
    European Congress on Biotechnology (2005), 53–56
    International Union of Microbiological Societies (2002), 284–287
    Society for Applied Microbiology (2006), 74–78
    Society for General Microbiology (2007), 79–82
    Society of American Bacteriologists, *see* Society of American Bacteriologists
    World Congress of Biotechnology (2000), 57–60
Melnick, Joseph, 258, 260
Melville, David, on avian influenza, 103
Mendel, Gregor, 3–4
Merigan, Tom, 235, 260
Merril, Carl, on bacteriophages, 230
Metapneumovirus, 285
Metchnikoff, Elie, cholera experiments of, 238–239

Meteors, jelly ascribed to, 30–33
Methyl bromide, bacteria degrading, 23
3-Methylcrotonyl-coenzyme A carboxylase deficiency, 320
Meynell, Guy
    on art work deterioration and restoration, 13
    on foxing, 43–44
Microbacterium flavescens, in citrus juice debittering, 45
Microbe Hunters (de Kruif), 41, 220
Microbiology, identity of, 314–317
Micrococcus luteus, in stem cells, 145
Microcystis aeruginosa, in waterfowl die-off, 322–323
Microscopes, Leeuwenhoek construction of, 201–205
Middelboe, Mathias, on viruses, 81
Migration, of locusts, 8–12
Milk, pasteurization of, controversy over, 151–155
Mink farms, astroviruses in, 286
Mirage of Health (Dubos), 207, 215–219, 239
Misra, Peter, on botulinum toxin, 37
Modern Science and the Nature of Life (Beck), 224
Molin, Soren, on biofilms, 74–75
More, Henry, on star jelly, 31
Moritella, in aquaculture, 272–273
Mosquitoes, as virus vectors, 284–285
Moxon, Richard, on virulence, 129
Mudd, Stuart, 258
Muirhead, Richard, on *Escherichia coli* in cowpats, 266–268
Multiple carboxylase deficiency, 320
Muto, Manabu, on acetaldehyde generation, 169
Mycobacterium avium, in protozoa, 123
Mycobacterium marinum, in lizards, 142–143
Mycobionts, lichen association with, 18–20
Myxomatosis, 175–178
Myxomycetes, in star jelly, 32

Paulino, Marianna, on cigar manufacture, 53–54
Pawelczyk, Adam, on bioremediation of pollution, 60
Peak, Nicholas, birds as disease vectors, 4–5
Penicillium, in cecina, 44–45
Peptic ulcer disease, *Helicobacter pylori* in, 83–86
Pérez-Ortin, José, on wine yeasts, 46
Pesticides, genetically modified, 133–137
Petroleum spills, bioremediation of, 296–299
Pettenkofer, Max von, 238–241
Petty, Nicola, on viruses, 81
Phages
    component transfer of, 278
    discovery of, 228–232
    periodic variations in, 80–81
Phenol, in locust swarming, 8–12
Pheromones, in locust swarming, 8–12
Phipps, James, vaccination of, 114
Photobacterium fischeri, quorum sensing in, 276
Photobionts, lichen association with, 18–20
Pichia pastoris, in collagen synthesis, 306
Pigs
    antibiotic use in, 118
    foot-and-mouth disease in, 97–98
    new rotavirus in, 286
Pirie, N. W. "Bill," on new notions, 277–278
Pisa Monumental Cemetery, art restoration in, 13–16
Pittet, Didier, on natural disasters, 91–95
Plague, in literature, 39–40
Plankton, *Helicobacter pylori* in, 84–86
Plants, *see also* Agriculture
    antibiotic resistance in, 279–281
    Enterococcus in, 280–281
    fermentation of, in cigar making, 53–54
    new viruses in, 286
    protection of, bacterial L forms in, 295

Plotkin, Stanley, 260
Pneumonia
    in literature, 41–42
    severe community-acquired, unrecognized by media, 156
Poliomyelitis
    eradication of, 114, 116–117
    vaccination for, 171–174
Pollution
    bioremediation of, 60, 189–193
    from bird droppings, 4–6
    from cowpats, 266–268
    cyanotoxins in, 322–324
    Helicobacter pylori in, 84–86
    in stone building destruction, 15–16
Porter, Ian, on typhoid fever outbreak, 158
Pott, Sir Percival, on cancer causes, 225
Poultry, antibiotic use in, 118
Pramer, David, 258–260
Priest Pot, England, biodiversity in, 24–25
Priobes, 208
Probiotics, for fish infections, 271
Proteobacteria, N-acetyl-L-homoserine lactone of, 277
Protozoa
    microorganism evolution and, 122–125
    as pathogen reservoirs, 282
Pseudomonas aeruginosa
    in biofilms, 70–72
    environment-human transfer of, 297–299
    intercellular signaling in, 276
    isolates of, 296–299
Pseudomonas fluorescens, in biofilms, 72
Pseudomonas mendocina, 297–298
Pseudomonas putida, in biofilms, 74–75
Pseudomonas stutzeri, in fresco restoration, 14–15
Psittacosis, in veterinary teaching hospital, 143–144
Psychoneuroimmunology, 61–65

Animalcules: the Activities, Impacts, and Investigators of Microbes

Animalcules: the Activities, Impacts, and Investigators of Microbes

Sussman, Max, on *Escherichia coli,* 49–50
Swarming, of locusts, microorganism involvement in, 8–12
Szybalski, Waclaw, 260

T

Taihu (lake), China, cyanobacteria in, 324
Tavío, Maria, on antibiotic resistance, 88
Taxonomy, relevance of, 296–299
Teague, Oscar, 211
Teloschistes capensis, sexuality of, 18
Terrorism, P4 high-containment laboratories and, 288–289
Tetanus, neonatal, eradication of, 116–117
Theiler, Max, 210–214
Thevelein, Johan, on glucose receptors in yeast, 307–308
Thompson, Kimberly, on Lyme disease vaccine, 140–141
Thorson, Anna, on avian influenza, 267–268
Tick(s), carrying *Borrelia burgdorferi,* 138–141
Tick-borne encephalitis, 285
Tischler, Max, 260–261
Tobacco, fermentation of, 53–54
Tombusvirus, in plants, 286
Toole, Michael, on disasters, 92
Torrance, Alfred, on pathogen extermination, 215
Torres, Juan, on myxomatosis, 177–178
Totman, Richard, on psychoneuroimmunology, 61–64
Toxins
    Clostridium botulinum, 34–38
    cyanobacteria, 322–324
    Escherichia coli, 48–52
    scorpion, 133–137
Tracking Down Enemies of Man (Torrance), 215
Transplantation, 152–153
Tremella, in star jelly, 32
Trends, in research, 275–278

Tribe, Henry, on fungal degradation of cardboard, 26–29
2,4,6-Trichloroanisol, in wine, 44
Trichoderma, in corked wine, 44
Trypanosoma, Hoare work with, 246–249
The Trypanosomes of Mammals (Duggan), 248
Tsuchida, Takayasu, on bacterial isolation, 59
Tsunami of 2004, 91–95
Tuberculosis
    eradication of, 115
    pathology of, 207
    stress effects on, 64
Turner, Sarah, on quorum sensing, 277
Twort, Antony, 229
Twort, Frederick, 228–232
Typhoid fever outbreak, media coverage of, 157–160

U

Untersuchungen über Bacterien (Cohn), 221
Usnic acid, pharmaceutical properties of, 20
Usutu virus, 284–285

V

Vaccination
    diseases preventable by, 180
    epidemiology and, 100–104
    for exterminating pathogens, 114–117
    fish pathogens, 272–273
    foot-and-mouth disease, 96–99
    hepatitis B, 181–182
    Lyme disease, 140–141
    measles, mumps, rubella, 102–103
    myxomatosis, 175–178
    objections to, 171–174
    rationalizing, 179–183
    respiratory syncytial virus, 58
    yellow fever, 210–214
van West, Pieter, on fish pathogens, 270–271

Vandamme, Peter, on *Burkholderia cepacia*, 281–282

VanRooyen, Michael, on disasters, 94

Vegetables, washing of, 145

Veldkamp, Hans, 260

Verdoes, Jan, on carotenoid production, 306–307

Verrucosispora, discovery of, 25

Veterinary studies, *see also* Cattle; Pigs
    antibiotics as growth promoters, 118–121
    new viruses, 286
    opportunistic infections, 142–146
    student infections in, 142–144

Vibrio, in seawater, 86

Vibrio cholerae
    bacteriophage interactions with, 80–81
    early work on, 238–241

Vibrio vulnificus, phages of, 230

Villas-Boas, Silas, on *Clostridium proteoclasticum*, 54–55

Virulence, 126–129

Viruses, *see also specific viruses*
    cancer-causing, 225
    cyanobacterial symbiosis with, 23
    ecology of, 79–82
    genetically manipulated, 175–178
    in *Leishmania*, 286–287
    new, 284–286
    symbiosis of, 23

Vogel, Gretchen, on disasters, 92

W

Waksman, Byron, 260

Waksman, Selman
    on soil organisms, 257
    on unified approach to microbiology, 279, 282

Waldenström, Jonas, birds as disease vectors, 5–6

Waldman, Ronald, on disasters, 92

Walker, Malcolm, on genetically modified foods, 185

Wallis, Tim, on virulence, 129

Walrus, calicivirus in, 286

Wang, Chengshu, on scorpion toxins, 134

Warren, Robin, on *Helicobacter pylori*, 83–84

Wastewater treatment, biotechnology for, 58–60

Water pollution
    bioremediation of, 60
    from bird droppings, 4–6
    from cowpats, 266–268
    cyanotoxins in, 322–324
    Helicobacter pylori in, 84–86

Webb, Jeremy, on biofilms, 75–76

Weber, Roland, on fungal degradation of cardboard, 26–29

Weissenbock, Herbert, on Usutu virus, 284–285

Welch-Novy-Russell Lecture, 258–259

Wells, H. G., autobiography of, 42

West Nile virus, Usutu virus similar to, 284–285

Westlake, Donald, on *Pseudomonas aeruginosa* isolates, 297–299

White, Andrew, on star jelly, 30, 32

White, David, 260

Wiesel, Torsten, on Noguchi, 213

Wildy, Peter, 260

Wilkinson, Marjorie, on *Escherichia coli*, 50–51

Williams, Greer, on Noguchi, 212

Williams, Michael, on typhoid fever outbreak, 158

Wilson, Mary, on transmitting pathogens, 218

Wine
    corked, 43–47
    yeasts for, 46–47

Winogradsky, Serge, Theiler association with, 255

Winslow, Charles-Edward Amory
    on pathogen extermination, 215
    Welch-Novy-Russell Lecture, 258–259

Wolter, Friedrich, as anticontagionist, 239–241

Wood, Thomas, on biofilms, 72

Woodmansey, Emma, on biofilms, 77

World Congress of Biotechnology, meeting of 2000, 57–60

World Health Organization, disease eradication programs of, 115–117, 173

X

Xanthophyllomyces dendrorhous, in carotenoid production, 306–307

Xanthoria parietina, sexuality of, 20

Y

Yamagiwa, Katsusaburo, on cancer causes, 226

Yamakawa, K., on biotin, 321

Yaws, eradication of, 116

Yeasts

complex nature of, 300–304

practical applications of, 305–308

wine, 46–47

Yellow fever

eradication of, 115

vaccine for, 210–214

Yersinia pestis, in literature, 39–40

Yuan, Xunlai, on lichen fossils, 20–21

Z

Zaehner, Hans, on antibiotic development, 294–295

Zinsser, Hans, 211

Zook, Douglas, promoting scientific literacy, 163